HOW INFRASTRUCTURE WORKS

HOW INFRASTRUCTURE WORKS

Inside the Systems

That Shape Our World

DEB CHACHRA

Riverhead Books

New York

2023

RIVERHEAD BOOKS
An imprint of Penguin Random House LLC
penguinrandomhouse.com

Copyright © 2023 by Debbie Chachra
Penguin Random House supports copyright. Copyright fuels creativity,
encourages diverse voices, promotes free speech, and creates a vibrant culture.
Thank you for buying an authorized edition of this book and for complying with
copyright laws by not reproducing, scanning, or distributing any part of it in any
form without permission. You are supporting writers and allowing Penguin
Random House to continue to publish books for every reader.

Riverhead and the R colophon are registered trademarks
of Penguin Random House LLC.

Library of Congress Cataloging-in-Publication Data
Names: Chachra, Deb, author.
Title: How infrastructure works: inside the systems
that shape our world / Deb Chachra.
Description: New York: Riverhead Books, 2023. |
Includes bibliographical references and index.
Identifiers: LCCN 2023014882 (print) | LCCN 2023014883 (ebook) |
ISBN 9780593086599 (hardcover) | ISBN 9780593086612 (ebook)
Subjects: LCSH: Public works. | Municipal engineering. |
Infrastructure (Economics)
Classification: LCC TA148 .C44 2023 (print) |
LCC TA148 (ebook) | DDC 363—dc23/eng/20230728
LC record available at https://lccn.loc.gov/2023014882
LC ebook record available at https://lccn.loc.gov/2023014883

Printed in the United States of America
1st Printing

Book design by Amanda Dewey

Dedicated to all those individuals—
past, present, and future—
whose ongoing labor, care, and attention
make our essential infrastructural systems possible.

CONTENTS

Central to any new order that can shape and direct technology and human destiny will be a renewed emphasis on the concept of justice.

Ursula M. Franklin, The Real World of Technology

HOW INFRASTRUCTURE WORKS

One

BEHIND THE LIGHTS

On a rainy Mother's Day in 2002, along with two hundred thousand other people, I went for a walk on a brand-new bridge.

The Leonard P. Zakim Bunker Hill Memorial Bridge in Boston is brilliant white, all soaring towers and parallel lines, echoing a nearby obelisk and evoking a ship under full sail. It's one of the world's widest cable-stayed bridges, and its roadbed carries ten lanes of Interstate 93 traffic over the Charles River before diving under the peninsula of the downtown core. A few weeks before it was opened to cars, local residents were invited to cross it on foot. And we came, five times as many visitors as were expected, despite the weather. My friend and I waited for hours, huddled under our umbrellas, before we finally made it onto the bridge. We craned our necks to drink it all in, knowing it was almost certainly the one and only time we would get to traverse it at our own speed and under our own power.

In retrospect, the organizers probably shouldn't have been quite so surprised at the demand. In 1987, when the Golden Gate Bridge was fully opened to pedestrians to celebrate its fiftieth anniversary, nearly three hundred thousand people were on the deck—more weight than bumper-to-bumper traffic—and the organizers had to turn away another half a million.

Conservation biologists sometimes refer to animals like pandas, elephants, and polar bears as "charismatic megafauna." Large, easily recognizable, and beloved, they make excellent spokescreatures for wildlife and for their homes. All over the world, charismatic megastructures are the public faces of our collective infrastructural systems. Bridges, of course, but also the soaring spaces of railway stations, from Grand Central Terminal in New York City to the Chhatrapati Shivaji Maharaj Terminus in Mumbai. In my hometown of Toronto, the CN Tower, built to be a communications hub, makes the skyline instantly recognizable. The Hoover Dam, in the desert outside Las Vegas, hosts seven million visitors every year. We *love* our charismatic megastructures. Their size and complexity might inspire awe, but I think it also has to do with the way that we feel like they belong to all of us, not individually but together, in much the same way as cathedrals and other sacred spaces. Not for nothing did Joan Didion describe driving on Los Angeles highways as a "secular communion."

Years after I dragged my friend out in the rain with me to walk on the Zakim Bridge, my nephews came to visit and we took an amphibious boat tour of the city. The guide told us about that one, now-legendary day in May when the bridge was open to pedestrians. I remember, I told them. I was there. My family was unsurprised, because they know I've been fascinated by science and engineering my entire life. The first thing I ever remember wanting to be as a kid was an astronaut—it was the 1970s, the cultural peak of space exploration. By the time I was ten, I wanted to be a nuclear physicist, and this took me all the way through to a degree in engineering physics. Like many women of my generation who went into the field, I had a role model close by, my father. He'd grown up and earned a degree in electrical engineering in a newly independent India before deciding he wanted to see more of the world. In the late 1960s, Canada, like the U.S., was opening its doors to non-European immigrants. My father enrolled in an MBA program at the University of Alberta and when he graduated, he took a job with Ontario Hydro, the public electrical utility for the province. His young family moved to the outskirts of Toronto and he started work at the place where he would eventually spend his entire professional career. A few months later, I was born. And a few months after that, in the next town over, one of the first large-scale nuclear power stations in Canada came online. I grew up on a street that ran down to a sandy beach on Lake Ontario, and when we picnicked and played in the water, the plant was easily visible, eight imposing gray concrete domes and a single massive cylindrical structure on the shoreline. Behind the facility, rows of electrical transmission towers marched along a wide, green right-of-way, first through the nearby suburban housing and then all over the city and region. Best of all, for my tech-obsessed childhood self, they had a visitor center with a nuclear power–themed interactive science museum. In my memory, it's all buttons and flashing lights, with cutaway models explaining the different systems and a walk-through mock-up of the reactor core. The Pickering Nuclear Generating Station was my local, personal charismatic megastructure. When I had to give a five-minute speech to my sixth-grade class at school, my topic was obvious: my fellow students got a lecture about the CANDU (CANadian Deuterium Uranium) reactors at our local power station, illustrated with a set of overhead transparencies my dad had brought home from work.

Scientists and engineers have a concept of *elegance*, which is among the highest praise that can be given to a theory, a proof, or a device. It's hard to explain exactly what it is, but it incorporates ideas of efficiency or parsimony—accomplishing something with a minimum of effort, energy, or materials. Elegance in engineering often encompasses the surprising and very particular use of the specific resources that are available. Beauty might be part of it, but mostly as something inherent in the design, and the skill and care of its manufacture is part of it too. For example, the graceful curves of suspension bridges, like the Brooklyn Bridge or the Golden Gate, derive from the way that cables are shaped by gravity as they bear the weight of the roadbed below. My local power station probably wouldn't be described as beautiful, but what was going on inside possessed undeniable engineering elegance. The reactor systems are carefully designed to produce and control the nuclear reactions that create heat, turning water into the pressurized steam which spins turbines and generates electricity. The CANDU design uses deuterium, an isotope of hydrogen, in the form of naturally occurring heavy water that's been purified out from ordinary water. It plays a safety-conscious dual role: the heavy water bathes the reactor core, supporting the nuclear reactions and also serving as a coolant. Because the reactor needs the deuterium to remain critical, a leak means that the core will fizzle out instead of overheating. What's more, the deuterium is so effective at sustaining nuclear fission that the uranium fuel only needs to be lightly enriched. That makes it easier and less dangerous to produce than other reactor fuels, and it's also much less amenable to being repurposed to make nuclear weapons.

Of course, as a kid I didn't think about any of the larger issues like limiting the risk of nuclear isotopes falling into the wrong hands or ways to deal with the long-term storage of radioactive waste. But even then, years before I took my first physics or engineering course, I could recognize the elegance and utility of the reactor design and how it made the most of what was available, whether that was locally mined uranium in the reactor core or frigid water from the depths of Lake Ontario for cooling. Today, I recognize this same elegance all over the world and across infrastructural systems, in a giant battery in Wales that's constructed from two lakes and the mountain between them and in the modular design of a solar plant outside Hyderabad, India. That elegance is a hallmark of infrastructure because these systems are almost always designed around the effective use of resources—above all, the use of energy. For most of human history, access to energy has enabled or limited what's possible. Harnessing and delivering energy, efficiently and at scale, has long been one of the main drivers for why and how societies have built out collective infrastructural systems.

The Zakim Bridge became an iconic part of the Boston skyline as soon as it was constructed. There are practical reasons for it to be such a dramatic structure—instead of being built on piers, it's a cable-stayed design, which limits its impact on the Charles River below—but surely it was also so that a monumental effort to make something *less* visible would be recognized and valued. The bridge was built as part of the Central Artery/Tunnel Project, better known as the Big Dig, the decades-long, multibillion-dollar project in which the elevated expressway that carried Interstate 93 across downtown Boston was torn down and the highway was buried underground. Elephants, whales, and other charismatic megafauna help us to see and appreciate entire ecosystems, including all of the unregarded but critically important species, like tiny insects and nondescript flora. Charismatic megastructures can similarly help us see and appreciate whole infrastructural systems, even the parts that are buried or hidden. And while I might have started with my massive local nuclear power plant, I began to realize that I could see and learn from the tiniest parts of these networks too.

Seeing Beneath the Surface

The Old State House in downtown Boston was built in 1713, making it one of the oldest public buildings in the city. I must have walked past it dozens of times before my eye was caught by the bronze glint of a benchmark sunk into the granite steps of a side door. It's a medallion, a few inches in diameter, with markings that indicate that it was placed by the U.S. Coast and Geodetic Survey to serve as a reference point for mapping. I had to kneel and look carefully at its worn face in order to make out the date: 1936. Similar survey markers can be found all over the world—I once startled a friend by spotting a tiny red marker, labeled CADASTRE, as I was stepping off a curb in Lausanne, Switzerland (it would have been used to create a cadastral map, an administrative record of all the real estate in the country). A few years ago, I went to the southern end of Central Park to find a small, anonymous steel rod sunk into a granite boulder. It was a Randel bolt, one of the markers that were used to lay out the original grid pattern of Manhattan streets at the start of the nineteenth century. On the modern suburban campus where I teach engineering, there's a small, bright blue metal pin embedded into a pathway to mark where a nineteenth-century aqueduct passes under the athletic fields. In urban centers, cryptic symbols spray-painted onto the pavement tell utility workers where to dig and what they'll find when they do, in much the same way doctors use surgical markers to make notations on patients before making the first incision in an operation. Boston is one of a number of cities that uses repair tags, plastic disks that are labeled with the name of the utility company and a reference number. They're embedded into asphalt repair patch after work is complete in order to let the next work crew know that something has changed. On unpaved ground, I've spotted flags or feathery bundles of colored plastic fiber used to mark where gas and water lines lie below the dirt. The bright colors aren't just for visibility. They're a code: in most of the U.S., blue is used for potable water, green for sewage and runoff, red for electricity, orange for communications, and yellow for natural gas, with hot pink used for temporary markings.

At any given moment, there is almost certainly some element of some infrastructural system in my field of view. Indoors, I can look around to see electrical lights and outlets, and usually there's a heating vent, and maybe a faucet or a drain. A few tiny curved lines in the corner of my laptop screen or straight lines on my phone tell me I'm on a global telecommunications network. Outdoors, there's almost certainly a nearby road, or at least a footpath that connects to one. But seeing isn't perceiving—just because something is in our field of view doesn't mean that we consciously register it, much less understand what we're looking at. Humans are disturbingly good at filtering out anything in our vision that we're either accustomed to seeing or which doesn't appear meaningful. Charismatic megastructures notwithstanding, many of the visible parts of infrastructural systems fall into this perceptual chasm. Between their familiarity and their unannounced, unexplained presence, infrastructural systems are easy to see but just as easy to ignore, unless we bring our conscious attention to bear on them.

Once we do, we begin to see that we are surrounded by networks, made visible not just by colorful squiggles but by sewer grates, telephone poles, mailboxes, access covers, transformer boxes, fire hydrants, railway lines, and on and on and on. The contrail of a jet airplane makes the invisible air travel corridors in the sky above us temporarily visible, just as spray paint does for underground ducts. Survey markers and street signs remind us that all of these networks were made possible by the centuries-long project of measuring the world, first with maps, then with clocks, and now with modern satellite mapping, which entangles the two. Even if the networks themselves are buried or intangible, we can visualize them by connecting the traces that we can see, knitting together people, buildings, and the world around us, in our home neighborhood or across the planet.

These visible markers show us not only the extent of these networks in space but also how they extend through *time*. Every marker, whether it's an imposing monument or a scrap of fabric, is there to send a message from the past to the future. Infrastructural networks are enduring, and that endurance shapes the way that new systems build on top of—or underneath, or alongside—the older ones. Water mains and gas pipes run along and beneath roads. Train tracks often followed watercourses as a path of least resistance through the terrain, telegraph lines were sent alongside the rails, and then roads were built along the same path, and electrical wires strung, and telephone cables. Because these networks are so important to the ways we do things, they have unparalleled continuity. We almost never rip them out and start over from scratch. Instead, infrastructural systems are repaired or rebuilt in modular increments, like steadily working through the replacement of water mains in a neighborhood or fixing potholes every spring. But that continuity has a flip side: it locks us into these ways of doing things. The QWERTY keyboard layout was designed to keep fast typists from jamming the keys on manual typewriters, but it's still in use on smartphone touchscreens. Decisions made decades or even centuries ago—how we treat wastewater, the use of alternating current instead of direct current for electricity grids, pipelines laid for fossil fuels—all of these shape not just the technologies and systems in use today but those that haven't yet been built. That continuity means there's a path dependence—that the kinds of systems we have today depend on the characteristics of the systems that came before—in addition to growth and accumulation, as these systems build on each other. We now live surrounded by technological systems of nearly unimaginable scale, extent, and complexity.

Landscapes and Cloudscapes

My family's immigration story is intertwined with the rise of global civil aviation, and so my childhood was one of transcontinental flights to India to visit extended family. Even after a lifetime of flying, it's still a thrill for me to get on a plane, to feel the press of acceleration during takeoff and the lift as the wheels release their contact with the ground, and to gaze out of the window for hours, at the sky or the landscape or the ocean below. Air travel has never become mundane to me, not least because once I think about what's happening, it's a *lot*. First of all, there's the sheer improbability of being in a vehicle that weighs hundreds of tons and is somehow still miles above the ground, airborne only by virtue of its tremendous forward speed. Along with this is understanding that it's because of the expertise of the pilots and the continued functioning of the massive jet engines, along with the systems that connect them to the control surfaces of the plane, to say nothing of the ones that keep the cabin comfortable in the face of the killing cold and the suffocating absence of air just on the other side of the porthole, a few inches from my face.

Those are just the most obvious, visible parts. When we're over the middle of the Atlantic, quietly seated, or asleep or watching a movie, we're all comfortable and the engines continue to roar quietly because of everything that happened before we boarded: the inspection, maintenance, and repair of the plane, and its provisioning with supplies, from aviation kerosene to toilet paper. And behind that, all of the standards, regulations, and certifications, everything from the fire resistance of seat fabric to the rules around how many hours the flight crew can work, from how and when to file a flight plan to the pictographs in the toilet. Then beyond *that*, an entire international system of atomic clocks and satellites so that all the planes in flight can agree on exactly what time it is and where they are, as well as the global network of weather sensors and prediction. Very few of these technologies and systems even existed a century ago, and now they're far too complex for a single mind to grasp.

Carl Sagan famously said, "If you wish to make an apple pie from scratch, first you must invent the universe." Ever since I was a kid, whenever I find myself someplace far from city lights on a clear night, I take some time and stare up at the stars. It reminds me that we're all whirling through space, and that everything I can see in the sky will continue to move along its own path regardless of anything that happens on our planet. Sometimes it can feel like cosmic vertigo, but I mostly find the vastness and indifference of the universe strangely reassuring. Because if you wish to drink a Bloody Mary and watch a movie as you fly somewhere for a vacation, first you must invent an unutterably and increasingly complex international system that's not just technological but social and political and regulatory, in order to make an intercontinental flight safe and even mundane. When I think too hard about air travel, I can give myself the same vertiginous feeling I get when I stare up into a starry sky, but I find being enfolded in these global systems similarly reassuring. It's helped by the ordinary human reassurance that comes from observing or interacting with professionals working with competence and care. It's part of why I like it when planes have an in-flight channel for the radio chatter from the cockpit.

But here's the kicker—or rather, two kickers. One is that, unlike the stars, the systems that enable my transatlantic flights are *human* systems, created and sustained by and for humans. They were designed and built to get at a certain set of outcomes—safe air travel, for sure, but also profits for airlines, or to bolster national pride, or to provide mobility options for residents of that country, or other reasons. The other kicker is that even if you're firmly on the ground, if you're reading these words, you are almost certainly still enmeshed in modern globe-spanning technological systems that have at least as much scale and complexity as civil aviation. The last half century or so has seen the rise of planetary networks not just for the movement of people but also of goods, in the form of supply chains enabled by the standardized shipping container, and of information, enabled by modern telecommunications. The growth in scale and extent of these systems have been in parallel with a steady increase in technological complexity, present not only in the obvious examples like computers but also in seemingly simple artifacts, like flimsy disposable plastic shopping bags. Whether it's a telephone or a T-shirt, most modern goods are made possible by cooperation and standards, the products of humans working together to make use of technologies that no one person can understand in their entirety. Every light switch connects to a system that's just as complicated as civil aviation and just as much a human creation full of human actors, as dependent on schedules, maps, standards, and regulations. The electrical grid, municipal water supplies, waste treatment facilities, motorized transport, food production systems, fossil fuel distribution, telecommunications—these technologies and networks underpin every aspect of the lives of a large fraction of the humans on the planet, whether directly or indirectly.

The Systems of the World

I've referred to a number of different systems, individually and collectively, as infrastructure. We can certainly intuit that they have something in common, so that it makes sense to group them together, but it's hard to articulate exactly what that is. What makes infrastructure, infrastructure?

"All of the stuff that you don't think about" turns out to be a surprisingly good starting point. For something to be considered infrastructure, its presence and characteristics are taken as a given. My bedside lamp needs electricity to function, and it plugs into an outlet, but I don't have to think much about the specific characteristics of the electricity: the voltage, the current, the frequency. The washing machine I use for my clothes needs not only electricity but also a supply of clean water. Some of that water comes from a hot-water heater, which itself relies on a piped-in supply of natural gas, and the dirty water drains into a sewage line. Four different infrastructural systems converge when I do my laundry. My ability to take these infrastructural systems for granted implies that there are underlying social agreements about what systems will be present and how they will function. They're not the only systems like this in our lives. A monetary system is a social agreement that money has value and can be traded for useful goods and services. A well-functioning legal system is a collective system that, among other characteristics, allows strangers to enter into contracts with confidence because all parties know that there are consequences for breaking them.

What electricity, water and sewage, telecommunications, and transportation—these familiar infrastructural systems—also have in common with each other is that they're technological systems. They're rooted in physical phenomena—the characteristics and behavior of matter and energy—and so require some kind of engineering to harness those phenomena. It might be as simple as digging a ditch to channel water flowing downhill into a slightly different direction, or it might be as complex as all the multitudinous technologies, from metallurgy to meteorology, that go into building a commercial jet plane and keeping it safely in the air. As an engineering professor, especially given my background in materials science, I have a deep awareness of how technologies are embedded in the physical world and how this makes infrastructure different from legal or monetary systems. Buying a cup of coffee might incorporate all sorts of technologies, from papermaking to RFID readers, but money itself doesn't require technology, much less any specific technology, to work. All it requires is that social agreement.

While infrastructural systems do involve social agreements around technologies, they share another important element. It's not just that every-one does things the same way. It's that there's a reason to do things the same way, together.

Most of the technological systems that we consider to be infrastructure are networks, connections between different places in the world. This is most obvious for transportation, ways to get from point A to point B: roads, rail networks, air travel, shipping, or transmodal combinations. Faucets and drains are points on our water and sewage systems that are connected by water mains, sewage pipes, reservoirs, and treatment plants; the oven in my kitchen is a node on the municipal network of natural gas pipes, as is the furnace of the building where I live. We're also embedded in electrical net-works that light up and power our homes and workplaces, and telecommuni-cations networks that connect us to others, once by dots and dashes, then by analog signals, and now increasingly by digital data, in all its varied permu-tations.

What makes networks special is not just that they're collective but that they can also be synergistic: the more people who use them, the more valu-able they become for everyone. The larger the population of a settlement, for example, the more incentive there is to make the substantive upfront in-vestment in building out a shared water supply. Every community, home, or business that is connected to a road or a telephone network makes the whole network more valuable for everyone else on it. The more people in your neighborhood who use electricity, the cheaper it becomes to deliver it. What's more, infrastructural systems often have benefits that go beyond the individual. Access to clean water reduces the risk of contagious waterborne disease for everyone, and electric lighting reduces the risk of fires. This intersection—technology, networks, and the value of universal access—is the key to why infrastructure is often considered a public good, in both the economic and everyday sense. And it's not static: the development and adop-tion of new networked technologies leads to new ideas of what's considered to be a universal need. We've been seeing this evolution happen in real time over the past decade or so as home broadband Internet access went from being useful but nonessential to being a primary means of interacting with systems and public services, including education.

As a child, I was most impressed by the technological elegance of infrastructure, and I still think that these systems are fascinating. As an adult, though, I've begun to see the networked, technological systems that are the focus of this book in a much different way. We might interact with them as individuals but they're inherently collective, social, and spatial. Because they bring resources to where they're used, they create enduring relationships not just between the people who share the network but also between those people and *place*, where they are in the world and the landscape the network traverses. These systems make manifest our ability to cooperate to meet universal needs and care for each other. We need those skills and behaviors, maybe more than ever before in human history. But that cooperation and care is by no means the whole story, because these systems also fail many people and even cause harm. Our infrastructural systems tell a story of who we are as a society or even a civilization, one that's about the relationships we have with one another and the planet that's our home. Hearing that story is part of learning how we can do better, and why it matters to all of us.

Situated Perspectives

When I was nine, my family lived in Bhopal, India, in my father's family home, for six months.

If culture is everything that you do without thinking much about why you're doing it, then our infrastructural systems, and the ways of life they make possible, are unquestionably an important part of culture. Even as a child, the differences between Canada and India—language, social norms, the deep poverty and the unignorable inequality, having strict Catholic nuns as teachers—required serious adjustment. There was also a culture shock associated with the infrastructural systems. The municipal water provision where we lived was on a rotating schedule, so we only had running water for an hour or so in the morning and again in the evening, and we collected it in buckets to use for bathing and flushing toilets all the rest of the time. My mother boiled and filtered the water to make it potable to digestive and immune systems that were accustomed to clean, cold, carefully treated water from Lake Ontario. We quickly learned to expect brownouts and power cuts as the growing city's electrical grid struggled to cope with the fans and evaporative coolers that were brought to bear against the summer heat.

My family moved back to Toronto in late autumn that year, and then I spent another six or so months in India when I was sixteen. That time in India made me unavoidably aware of the systems that were all working, quietly and continuously and reliably, to meet my basic needs as I was growing up. I doubt I would have noticed or thought anywhere near as much about infrastructure had I not lived in these two different places. By moving to Canada, my parents had given me a new citizenship in a country with a different set of educational and economic opportunities, alongside of which was the infrastructural birthright that underpinned my ability to access them.

Collective infrastructures are a good candidate for the most complex systems created by humans. They are planetary in scale, build on their own histories, are entangled with each other, and have impacts that extend far into the future. Their design, construction, and operation require a wide range of technical disciplines—civil engineering, obviously, but also electrical engineering, mechanical engineering, environmental engineering, and the science of systems and of networks. All of these fields incorporate not just technologies but practices, ways of thinking, doing, and building. The decision-making and funding around these systems, particularly public systems, is the province of policy, finance, and economics. Finally, as systems that require land and resources, it's impossible to extricate infrastructure from a larger history of conquest, colonialism, and capitalism and, particularly but by no means exclusively in the U.S., from racism and white supremacy. The theorist Timothy Morton coined the word "hyperobject" for just these sorts of systems that span multiple time scales, length scales, and social settings.

My friend Helen Macdonald's writing about the natural environment has influenced how I think about infrastructural systems. A forest is home to trees, plants, fungi, and lichen, as well as to birds, insects, reptiles, amphibians, small and large mammals, and more, each of which is in an ongoing ecological relationship with the others. Every forest is a hyperobject, an enormously complex environment that's shaped not just by its location, landscape, and climate but also by the history of humans in that place. If you go on a walk in the woods, what you see depends on the season and the particular path you take through it. More than anything, though, what you see in the woods depends on the eyes that you are seeing it through. A birder, a hunter, an entomologist, a soil ecologist, a real estate developer, and an artist will all see different things. Helen introduced me to Richard Mabey's idea that the natural world can only be understood and appreciated as a result of careful, knowledgeable attention. Without the ability to describe the differences with detail and specificity, a meadow of native grasses and the pesticide-soaked monoculture of a golf course that replaces it are effectively indistinguishable. Helen borrowed Mabey's phrase for this in an interview, calling it a "green blur." We'll be taking a closer look at the "gray blur" of our infrastructural landscape as we go on a walk through it, in the hope of resolving some of the details. But, as with a forest, it's just one trail with a particular guide, and what we see depends a lot on who's doing the looking.

Infrastructure is inextricable from culture, woven as it is into our daily life as well as our history, politics, civic life, and our visions of the future for our families and our communities. My path through infrastructural systems, and therefore what I see when I look at them, is deeply rooted in my own experiences. I've lived most of my life in cities in four countries: Canada, the United States, India, and the United Kingdom. I joke that as a dual

U.S.-Canadian citizen of Indian descent, I'm a colonial three times over. While that gives me a very specific perspective on the British Empire and its infrastructural projects, growing up in anglophone North America means that England has, in turn, shaped my perspective on the world. I also spent most of my adolescence and young adulthood training as an engineer and a scientist before becoming an engineering professor. It's how I came to the world of infrastructure, and it's a big part of how I think about and describe it: as networks and systems, energy and material flows, standards and technologies, efficiency and elegance, with a similar vocabulary and perspective as the engineers who built these systems out.

But it's at least as important to my perspective on infrastructural systems that I'm the daughter of immigrants who came from a postcolonial country in the Global South and settled in a rich, industrialized country. I recognize that I'm a historical anomaly, and something of an anomaly even today. All over the planet, the responsibility for hygiene and food preparation most often falls to women. The highly gendered nature of this labor means that fetching clean drinking water or obtaining fuel to burn for cooking are daily activities for billions of women globally, especially ones who look like me, brown-skinned and middle-aged. They trade their time and the energy of their own bodies to ensure that the most basic needs of their families are met. My life, as I know it, has mostly been made possible because I had the sheer blind luck of being born into a time and place where my survival needs were met and continue to be met every day by shared infrastructure. What might our world be like if access to these systems were universal?

Cracks in the Technological Foundations

Like most living creatures, humans habituate to whatever's unchanging in our environments. The more robust and reliable our infrastructural systems are, the less we think about them, and the more time and attention we put into other things. When I get up in the morning, I don't think about the water coming out of the bathroom tap in the same way that I don't think about the sun rising. But unlike the sun rising, our infrastructural systems are built by humans, for humans, and all technological systems need care and maintenance. Infrastructural systems are famously boring because the best possible outcome is nothing happening, or at least nothing unexpected or untoward. But nothing happens, and nothing continues to happen, as a result of sufficient attention, specialized care, and unceasing oversight. "Nothing happening" is usually the result of careful inspection schedules, preventive maintenance, and planned replacement, all of which require resources to be devoted to what will be, at best, a null outcome. Across the U.S., decades of underinvestment in infrastructure mean that systems like transportation and the electrical grid are failing more frequently, whether small failures like delayed subway trains, potholes, or a local power outage, or catastrophic ones like a bridge collapse or a multiday blackout. Even systems that might have been adequate when they were built are failing to keep pace with changes in the larger world, whether predictable ones like population growth or new challenges like contaminants in water supplies.

Failure is relative to expectations, of course. Access to a suite of reliable, modern infrastructural systems is the global exception, not the rule. Communities all over the world, including in the U.S., have unreliable or limited access to electricity, to clean water or improved sanitation, to energy for cooking, or to reliable telecommunications. Many decades after they were first built, we have a much more widespread understanding of the injustices that can be encoded in these systems—where they are built and for whom, where they draw resources from, and who's most impacted by negative consequences, from air pollution to flooding to noise and congestion.

But the biggest threats to infrastructural systems don't come from neglect or underfunding. They are inherent. Infrastructural networks are built to move resources from place to place—atoms and bits, electrons and photons, people and things—and to do so reliably and continuously. The substrate that they are embedded in and traverse is the landscape. Reliability and continuity presuppose a certain stability, but the climate is no longer stable. Rather than being something that we're headed toward, anthropogenic climate change is now something that we are living through. Globally, every year since 2015 has been among the warmest years ever recorded, and we are beginning to see the effects everywhere. The western U.S. has now seen years of record-setting drought, followed by years of intense rainfall and flooding, followed by years of deadly wildfires. In the eastern U.S., there was Hurricane Sandy and dark, flooded lower Manhattan. In Puerto Rico, communities were left without power for nearly a year after Hurricane Maria. In 2020, there were so many named hurricanes that we went through the entire Roman alphabet and started in with Greek letters, with Hurricane Zeta making landfall in the U.S. in late October. Every natural disaster is the result of an interaction between a natural event and human systems, and an important measure of the severity is the degree to which infrastructural systems are affected or how quickly they can be repaired. Functional infrastructure is resilient infrastructure.

All technological systems require energy to function, and the vast majority of that energy still comes from fossil fuels. Infrastructure is also a major *contributor* to climate change. In the U.S., electricity generation accounts for more than a quarter of all greenhouse gases. Transportation accounts for another 30 percent or so, and another few percent come directly from household use, like the natural gas that's burned to warm my home and cook my dinner. In total, energy use that's mediated by infrastructural systems accounts for close to two thirds of emissions.

What all of this means is that there is no more taking our infrastructural systems for granted, anywhere. In the Bugs Bunny cartoons, Wile E. Coyote races off the edge of a cliff, looks around for a moment, realizes there is nothing below him but thin air, and *then* he falls. If infrastructure becomes most visible when it fails, we are at the exact moment when, like

Coyote, we are looking around and seeing where our societal decisions have led us. But unlike Coyote's fall, there's nothing inevitable about what happens next.

Our Fractured Past and Our Shared Future

The hyperobjects that are our collective infrastructural systems shape not just what we actually do, day after day, as individuals and as communities, but even the possibilities for what we can consider doing. The cost of not being able to rely on previously reliable systems is unthinkably high: to have had well-functioning infrastructural systems and then to lose them is recognizably dystopian, the stuff of postapocalyptic science fiction. Which means that we can't live without them and we can't keep them as they are. What's more, we can only change them together.

But here's the thing: we have everything we need to create a different world, for everyone. We have myriad ways to harness renewable energy. It's possible to transform and rebuild our collective infrastructural systems to be functional and also resilient and sustainable. This gives us an unprecedented opportunity to rethink them for the coming century, instead of being defined by the previous one. That includes recognizing the fundamental injustices built into these systems, from the neighborhood to the planetary level, and redressing them. Fifty years ago, transforming these systems would have been an intractable engineering problem. No longer—today, the challenges we face are largely social and mostly boil down to this: How do we work together to transform these collective systems? Because the one thing we do know is that these shared systems that shape our lives, both affording and constraining the ways in which we do things, can't be changed appreciably by individual action alone.

My own relationship with infrastructure started when I was a young girl fascinated by huge engineering facilities, and I came to see and appreciate them as among the most compelling of human creations. Over the past decade or so, I went to stand at the edge of reservoirs, in a power station inside a mountain, on the crest of one massive dam in the desert, and on the ruins of another. I toured sewage treatment plants, walked a road that humans have followed for five millennia, stood between rows of tilted solar panels and among majestically spinning wind turbines, and walked the channel of a river through a city and down to the sea. These systems are how we care for our bodies wherever they are on the planet, and so I took my body to the systems to try to understand them. The more I learned, the more I saw how deeply our infrastructural systems were shaped by our cultures—often the best aspects of ourselves, as we committed to take care of each other, but also the very worst, literally building institutionalized racism into our highways, dams, and power systems, creating a world of haves and have-nots. I gradually began to apprehend all the ways our infrastructural systems don't serve us well—and to think more deeply about who counts as "us." I started to recognize the degree to which they are under threat, and I learned that we have the technologies to respond to these threats, but also that doing so lends itself to a very different way of organizing our communities and our society.

It might be a truism that you can't solve a problem with the same mindset that created it, but you do still have to first recognize and understand the problem and, however you approach it, the solutions need to be on the same *scale* as the problem. Our infrastructural systems make our individual daily lives possible, but they are vast and collective, and so they can only be seen, understood, and transformed collectively. This book is intended to be a stepping stone to that transformation, but it's not a how-to manual. It's a *why-to*, and a *where-to*, manual.

Two

INFRASTRUCTURE
AS AGENCY

On a typical evening, I might drive or take the subway home to my apartment. I let myself in, turn on the lights, put down my bags, and take off my coat and shoes. I turn the stereo on and set it to play music streamed from my phone, and I head into the kitchen to make dinner. If I'm cooking something like noodles for dinner, I can get a pot from the cupboard, fill it from the tap, and set it on the stovetop to boil, and then take some vegetables out of the fridge to prepare. After years of practice, I can usually be sitting down to a hot meal in less than half an hour. Once I've eaten, I take a few minutes more to clean up, putting the waste in the trash and washing the dishes in a sink of hot water.

When I was growing up, my parents had a small Hindu altar set up in our home, with mounted posters of several multiarmed gods and goddesses in the heightened-realism illustration style that's common for these icons. As a child, I was fascinated by the warrior-goddess Durga, who was depicted with weapons in many of her hands and riding a tiger that's trampling a demon under its paws. Despite what those terrifying attributes would suggest, Durga isn't associated with only war and destruction—she's also a goddess of protection. That's why she's almost always portrayed with a gentle, peaceful smile on her face. Durga Mata ("Mother Durga") fights not out of hatred or fear or even self-defense, but on behalf of those who depend on her, for their safety and liberation.

I described myself making dinner because it might be the quietest and most domestic part of my day. Let's get real: it's boring. But that very ordinariness is utterly extraordinary. Hiding in plain sight in this scene are many of the different ways in which our infrastructural systems augment what's possible for me to do with my body alone. Common infrastructural systems— water, sewage, electricity, natural gas, transportation, and communications— are amazing in that they make it possible for me to live the kind of life that I want to live, a life of comfort, agency, diversion, and connection to others. Some of them deliver useful matter, such as clean water, fuel to burn for useful heat, or physical goods. Another provides versatile energy, in the form of electricity. Infrastructural systems make it possible for my own body to move around the world so I can be with others, in other places. I can command fire with the flick of a wrist, even if I'm just going to use it to soft-boil an egg. On the coldest winter mornings, I wake up in a home that's warm and comfortable. Global telecommunications mean that I can stay in instantaneous connection with colleagues, friends, and family all over the planet. Riding my bike on paved roads takes me farther and faster than I could go on foot, or I could get into my car, or catch a train. And I can fly, crossing an ocean in hours. But even alone in my kitchen, I'm an infrastructure-goddess Durga, far seeing, with a host of tools and abilities at my fingertips. Above all, I'm *powerful*.

Since they're always there for my use, at any moment of the day, these infrastructural systems are all but part of me—they extend my capabilities beyond my biological limitations in a way that I almost never have to think about. I can't see in the dark, but abundant artificial light means that I might as well be able to. The interfaces to these systems are even scaled and shaped to my body: the intensity and color temperature of the lightbulbs, comfortably heated water for washing and bathing, and the ambient temperature of my apartment; the width of a sidewalk, the turnstiles at the subway station; the placement of power outlets; the size and shape of all the switches and knobs in my home; everything from my smartphone to the toilet.

The most direct example of this is my car, which essentially makes me into a *cyborg*, an actual, exoskeleton-wearing, Sigourney-Weaver-in-a-power-loader-fighting-off-the-xenomorph-queen cyborg. When I'm in the driver's seat, I'm wearing a metal shell that protects me, gives me equipment that doesn't exactly come standard on a human body (like wheels, sure, but also a stereo!) and, most importantly, gives me direct control over an enormous store of energy that is exogenous to my body itself. The car I've been driving has an internal combustion engine that produces about 150 horsepower, which makes it modestly powered by American vehicle standards but is still a considerable amount of energy for one person to be in charge of. On a full tank of gas, I can transport myself, four passengers, and a trunkful of gear about 350 miles (more than 550 kilometers) in about five hours. The utility of cars—their appeal—is exactly that they make us more powerful and capable as *individuals*. It's why the selling point of private vehicles, as featured in countless advertisements, has always been freedom: that you can choose when to leave and where to go, and your car will take you there. Cars serve as a personal interface to an infrastructural system of personal mobility: not just the network of roads, but also the fueling stations that are located alongside (there's somewhere upward of a hundred thousand gas stations in the U.S. alone), and the entire supply chain for extracting, refining, and delivering gasoline to all of them.

But here's the thing: when I flip a light switch, I'm just as much a human-machine hybrid as when I'm driving. The difference is that I'm standing in my kitchen, and the "machine" is the entire electrical grid—giant turbines spinning up to power huge generators, thousands of miles of power lines, the local substations, neighborhood transformers, plus the enormous amount of human expertise necessary to design, build, operate, and maintain these systems. Alone at home I am a continent-spanning co-lossus, reaching out for my tiny share of these communal systems, whether it's a few drops of water from my local reservoir via the municipal treatment plant or a few electrons getting a push from a dam in northern Quebec, thou-sands of miles away. If driving my car makes me a cyborg because it aug-ments my individual body with a more powerful machine, then these systems make me part of, in the words of infrastructure researcher Paul Graham Raven, a "cyborg collective."

A Brighter World

For many years, I've lived in an apartment in a four-story mixed-use building in central Cambridge, Massachusetts. It was originally built in the late nine-teenth century, constructed in an ornate Richardsonian Romanesque style, complete with arched windows, a tower, and elaborate carvings over the en-trances. I occasionally think about what the lives of the people who were in the space before me were like, but I also think about the converse: What if the original occupants of my building came a century and a half forward in time to my home in the present day—what would surprise them the most?

My personal library (an admittedly somewhat genteel way to say that my home is full of books) would certainly be impressive as well as unexpectedly colorful, thanks to modern printing technology. What I think of as a rela-tively modest personal wardrobe would come across as anything but to someone visiting from the early days of industrialization. It's almost all black and in plain cuts, which even nineteenth-century Bostonians might find oddly puritanical and old-fashioned if it weren't for the wide diversity of tex-tures, with wool and cotton hanging alongside modern synthetics and tech-nical materials. Although my visitors would be familiar with mechanized textile production—it was a major driver of the Industrial Revolution—the contents of my closets still represent an absolutely astounding amount of effort in spinning, weaving or knitting, and sewing. More generally, the sheer amount of *stuff* in my apartment might be overwhelming, especially in light of its precision and detail since almost everything I own is mass-produced by machinery rather than being made by hand. The refrigerator and freezer would be impressive but not astonishing, I suspect. The last decades of the nineteenth century were the peak of the natural ice trade, which employed tens of thousands of people in New England to cut, store, and ship winter ice, and artificial refrigeration technologies had already started to come into use for industrial applications. So my nomination for the most startlingly futuristic aspect of my home, especially on a winter's evening, would be just how *bright* it is, and how all of the light is cool and white, flameless and odorless, with no apparent fuel source and lit or extinguished with a flick of a switch.

It's easy to take artificial light for granted and never think about what an enabling technology it is to be able to see equally well whether it's day or night or, for that matter, in any closed interior space. It's not a survival need, but seeing in the dark is effectively a superpower. While light has been scarce and expensive for almost all of human history, the past two centuries or so saw the cost of light (the price per lumen-hour) drop by a factor of ten thousand. As Bostonians from the late nineteenth century, my hypothetical visitors would likely have been familiar with two important contributors to this dramatic change: the rise of coal-derived "town gas," distributed to buildings via municipal pipes; and the decline of the whaling industry and with it the use of whale oil for lighting, as kerosene from the rapidly expanding fossil fuel industry became the dominant liquid fuel for lamps. But even if they had heard about Thomas Edison in Menlo Park and his ongoing experiments with electricity and incandescent lightbulbs, they would probably still be unprepared for the brightly lit world that would result. Economic historians estimated that the total consumption of light in the UK today is about a *hundred thousand* times greater than it was in 1700. Winter days aren't quite as short here, but assuming a similar trajectory, modern New England uses about a hundred times more light now than in 1888. I can barely imagine what it would be like to live in that world, where it was mostly just dark, for months on end. In fact, access to more resources leads so universally and predictably to an increase in artificial light that researchers have even used it as a proxy for economic growth, observing how nighttime satellite images of a particular region glow more brightly over time. Developed regions are brighter than comparable regions in less developed areas. South Korea, for example, blazes from coast to coast, with a clear shoreline. But after years of sanctions and scarcity, North Korea is nearly invisible in nighttime images except for a faint cobweb of light reaching out from the capital of Pyongyang. Besides energy itself, barrierless access to artificial light might be the closest thing we have to pure agency, literally providing more hours in the day.

Warm Homes and Hot Dinners

Just as artificial light makes it possible for me to do what I'd like well into the night, living in a home with adequate heating means that I don't spend the winter huddled under blankets for warmth, or my days managing a stove. Like most urban buildings in the northeast corner of the United States, mine is heated by the combustion of natural gas, which is mostly methane with a few percent of ethane. The local distribution system started as the pipes that were originally built to deliver the town gas that my time travelers would recognize from streetlights (it's the gas in "gaslight"). The longer-range transmission network started as pipelines that were crash-built by a consortium of oil companies during World War II to bring Texas crude and oil products to New York and Pennsylvania for refining and transshipment to the European front. The construction of fifteen hundred miles of pipeline—the longest and biggest of its kind at the time—was mandated by the U.S. government and funded as a war measure because the sea routes were no longer safe or viable: barges loaded with oil and headed up the Atlantic coast were a soft target for German U-boats. Once the war ended, the pipelines were sold off as surplus, then retrofitted to carry natural gas and connected to the municipal networks. At the oilfields in the Southwest, large quantities of natural gas were an undesirable by-product of extraction, often just flared off at the wellhead. By getting it to buildings in the energy-hungry Northeast, oil companies could arbitrage and profit from the difference in value to the two regions. Once the pipelines made it possible to deliver natural gas efficiently (and therefore inexpensively), it quickly became the dominant fuel for heating in the region.

Besides powering the furnace and the hot-water heater in my building, some of that natural gas is delivered to the stove in my kitchen. Whenever I want to cook, I just turn a knob and the blue flame instantly whooshes out. Hot meals aren't a luxury; while individuals can survive without hot food, there are no known human cultures that don't cook their food. Raising the temperature of raw plant or animal matter causes chemical changes that makes the nutrients much more bioavailable—in other words, more nutritional value out of less foodstuffs. The species-wide uptake of this strategy speaks to its effectiveness, but it's also why obtaining fuel is a daily preoccupation for much of the world's population. For humans, eating means cooking. That's the global context for what it means for me to have all of the fuel I could ever need for food preparation at my fingertips. The local context, including the presence of a municipal distribution network and transmission pipelines and the federal and state policies that caused them to be built, means that the fuel that I use for food preparation costs me shockingly little in labor terms. My gas bill for a month of food preparation corresponds to only about an hour or so of work at the state minimum wage. Nearly as much as artificial light and potable water. Only needing to work a few minutes a day in order to have sufficient energy to cook all my food makes my life different from that of most women worldwide.

Mechanical Work and Physical Labor

After I've eaten, it's time to clean up and do other housekeeping. I don't have a dishwasher, so I wash my dishes by hand with the on-tap hot water that makes it fast and easy. In a rare and pleasing example of not needing to put energy into the system to get a desired outcome, if I set my dishes in the rack and leave them alone, they'll spontaneously dry themselves. I do use an electric washing machine for my laundry, a standard American-style top-loader, with a rotating perforated drum and a central agitator. There's no designated laundry day in my home—it gets started when I have a pile of dirty clothes and a spare ten minutes, and then I go do other things until the clothes are ready to go into the dryer, be hung up to air dry, or be folded and put away. I don't need to use my own physical labor to scrub sodden, heavy clothing to loosen dirt, nor to rinse it, nor to wring out the excess water. My washing machine runs on electricity, of course, but here that power is doing a very different job than it does in lightbulbs or in heating elements. It's being transformed into motion by means of a motor, producing the forces needed to spin the washer drum and to agitate my clothes inside it. Like a waterwheel, an electric motor produces rotary motion by default, but clever mechanical engineering can then get many other movements. Motors make it possible for electricity to replace all sorts of human physical effort—when we talk about labor-saving devices, this is usually what we're thinking about.

In the domestic sphere, much of this work would be performed with our own bodies, whether individually or in concert with others. This is ordinary human labor—moving ourselves around, moving things around—but our bodies are limited in important ways. The first is time. We can do things, sure, but there are only so many hours in a day and so many days in our lives, and the time spent in this labor is time *not* spent doing other things. Fetching water to drink or acquiring fuel for cooking are tasks that meet survival needs, and as such they take precedence over other activities. The second limiting factor in using the energy of our bodies to do work is that it results in fatigue. We get tired! Depending on the work we're doing, we might not have the physical stamina to do much else. Finally, powered machinery makes it possible to do things in ways that humans are otherwise incapable of, like vacuuming carpets instead of beating them to remove dust.

Everything I've described—the kinds of domestic labor I do, how I do it, and how often—is obviously very middle-class, early twenty-first century, and U.S.-centric. The nature of household work has varied enormously by time and place, and it's heavily shaped and constrained by social norms. In her groundbreaking 1985 work, *More Work for Mother,* the historian Ruth Schwartz Cowan argued that, despite the Industrial Revolution, the build-out of networks for fuel, clean water, and electricity, and the spread of household appliances, middle-class women were spending as much time on domestic labor in 1950 as a century earlier. Their work was industrialized, and they began to be held to higher standards like cleaning and doing laundry more often, or making more complicated meals. But even if they didn't replace work, what appliances did generally replace was *drudgery*—no more lugging water or coal, and a washing machine makes doing laundry a very different proposition from handwashing clothes. But even more saliently: in 1950, a housewife could single-handedly produce a middle-class standard of living for her family while her 1850 counterpart would likely have had one or many servants—"hired help"—to do the same. Those middle-class standards of hygiene and comfort (and the markedly reduced rates of illness that went with them) became widely available to many more people, not just to a small fraction of a population supported by the labor of the remainder.

In all of human history, only the smallest handful of people have been able to live the kind of life that I get to take for granted, and they've only been able to do so because others were doing all of the caregiving work that they needed. When we say that someone is "supported" by a partner, we usually mean that they're supported financially. We tend to lose sight of how domestic labor—doing work to make sure the people around you are warm, fed, clean, and healthy—supports survival. It's work that needs to happen every day, without fail. Globally, this is gendered labor, largely performed by women. But in my household, this work is eased by access to shared infrastructural networks that deliver the fuel whose combustion warms my home and cooks my food, clean water, and electricity that provides light and the mechanical energy to replace the physical labor of my body.

To Be with Others, in Many Ways

Humans move through the world. Our bipedal stance means that our dominant gait is walking—we swing our long legs like alternating pendulums, letting gravity do most of the work and catching ourselves at the end of each step. If we want to go faster, we can run, but running requires much more power since at every step we propel ourselves off the ground and forward. To go farther, faster, or move more weight, humans have used a range of strategies. One is to modify the surface and the path—even the narrowest trail means less resistance to forward movement. Devices like the wheel make motion more efficient so we can make the most of our limited bodily energy. Humans have also harnessed energy found in our environment, like flowing water or blowing wind, or even just an incline. If you've ever ridden a bicycle down a hill on a smooth road, you've used all three of these strategies put together. But the real step up in facilitating mobility is to bring more energy to bear. For many thousands of years, it was the power of other living creatures, from boisterous sled dogs to enormous draft horses. Now, of course, most of the energy for movement comes from technological sources, in the form of motorized transportation powered by electricity or by combustion. Personal cars might make us individual cyborgs, but motorized mass transit—buses, trains, ships, and above all aviation—puts energy to work to provide new and unprecedented access to personal mobility. Humans move around for the same reasons we've always moved around: to get to where we need to be to do work, to obtain something we need or desire, to explore our environment, and to move for the sheer joy of moving. But above all, we move around to be with other people, because sociality is another human universal. As transportation systems have grown in size, speed, and extent, so too have our communities, in both senses: the built environments in which we live, but also the local and extended groups of people with whom we have relationships. Personal mobility is enhanced by motorized transportation infrastructure with its continuing access to energy outside of our bodies. Our ability to be with the people around us has grown to encompass greater and greater distances, whether it's a daily twenty-mile commute or getting on a commercial airliner to fly to the other side of the planet in less than a day. These transportation systems are what allow me to be with my colleagues and my students on most days, and with my friends and family around the world.

While being physically copresent with others might be the oldest form of human connection, the newest infrastructural systems make it possible for us to stay connected with each other in ways that otherwise simply wouldn't exist. I live alone, and so my home is mostly about my not being in it. It's right at a subway station in a dense, walkable neighborhood; I drive a car and ride my bike; and I have quick, easy access to a train station, an airport, and even ferries. But for fifteen months or so—by far the longest stretch of my life—I mostly didn't access any of these transportation systems. I didn't take the T downtown, or the train to New York or Washington, nor did I get on a plane to visit my family in Los Angeles or outside Toronto. While the COVID-19 pandemic was a global public health crisis that touched the lives of everyone on the planet, the nature of its impact was widely variable. I was in the class of people who were largely able to shelter in place, because I have stable housing and a stable income and was able to depend on infrastructure, food, and other systems that were kept running by essential workers. But one thing that almost everyone, including me, had in common was that we limited the amount of time we spent with people outside our households in order to keep our communities safe. That meant that we collectively used transportation much less. It also meant that many of us used telecommunications—an entirely different way of staying connected—much more.

In February 2020, a few weeks before the college where I teach closed its doors and sent our students away, I was on the phone with my local cable company arranging to upgrade the minimal, janky Internet service that had always been all I needed because I spent so little time at home, and it had already become clear that I would soon be leaning hard on copresence via telecommunications. In a very real sense, my Internet connection replaced my transportation infrastructure for the duration. I taught my classes over Zoom; I stayed in touch with my friends and family through FaceTime and social media; I distracted myself with streaming video and tried to keep in shape with online workouts; and I did most of my personal administrative tasks, like banking and civic engagement, through websites. And honestly, it saved me. Not just in the practical sense of being able to fulfill my responsibilities as a college professor and pay my taxes, but because having a high-bandwidth Internet connection is what ameliorated the wrenching loneliness of getting through fifteen months of almost no physical contact with other human beings.

Global telecommunications enabled me to be, as anthropologist Caitrin Lynch put it, "physically distant but socially close" to her and my other colleagues, our students, and to the people in my life that I love. I'm old enough to not take the Internet, never mind videoconferencing, for granted. For most of my youth, the main way that my parents in Canada kept in touch with their families back home in India was by sending written letters through the post. There was the occasional card for birthdays or Diwali, but mostly the letters were envelope-less aerograms—single sheets of flimsy blue paper, folded into thirds, glued shut, and addressed and stamped on the outside.

For important news, there were telegrams—I'm just old enough to remember receiving telegrams from my extended family on my birthday. (The state telecom service in India was a long holdout, continuing to send general-purpose telegrams until 2013.) Long-distance voice calls—unreliable, expensive, and laggy—were reserved for the most urgent communications. I remember the long, sleepless night before my father succeeded in making contact with his brother and sisters in Bhopal as news of the Union Carbide industrial disaster began to trickle out.

As I was writing this, the periodic buzzing of my phone let me know that my siblings are posting to our family WhatsApp group, sharing photos of my niece and nephews and other snippets of their daily lives, and that my friends were sending messages ranging from cute cat photos to travel scheduling to political commentary. When my aging smartphone died unexpectedly on New Year's Eve a few years ago, getting a replacement suddenly became the top item on my to-do list for that day. My Internet connection was fine and so was the mobile phone network, but not having a working phone was a reminder that much of the value of infrastructural networks depends on being able to interface with the systems. Live electrical outlets in your home aren't useful without appliances that run on electricity, or ones with the wrong kind of plug. In that sense, my smartphone is like an astoundingly complicated multifunctional Internet-powered toaster. It's an appliance that makes it possible for me to interface with a network to do something useful, except that instead of making my bread crispy, it lets me use modern telecommunications to engage with online social infrastructure like the World Wide Web, messaging, and photosharing.

Telecommunications networks run, of course, on electricity. It powers all of the storage and computation behind every website and streaming service—large data centers have been sited near hydroelectricity plants to take advantage of inexpensive power—and also all of the connectivity and communication, the transmission towers, switching stations, and more.

Wireless communication, for example, is possible because electrical signals lend themselves to being converted to electromagnetic waves that can be sent through the air and then transformed back into electrical signals in wires, and that data can convey information in a wide variety of ways. If mechanized transportation is a way to use unprecedented amounts of energy to connect with each other in a way as old as humanity—by physically being with each other—then telecommunications is about using energy to connect with each other in entirely unprecedented ways.

Telecommunications systems illustrate another point, which is that we can be directly responsible for energy usage even if we never directly observe or control it. When I watch a movie on a streaming service, the electricity that powers the display and my Internet router—the electricity usage that shows up on my utility bill—is only part of the energy consumption story. It also requires data centers full of servers running twenty-four hours a day to store and serve up movies on demand, and the transmission of that data to my home. It's easy to never think about the energy usage of data centers that might be hundreds or thousands of miles away. But there's another way in which we are the end users of energy, one that's at least as important and ubiquitous yet even less obvious.

The Embodied Energy of Objects

For the past few years, I've been learning how to blow glass. About once a week, I head over to a glassblowing studio in an industrial building that sits on the railway tracks in north Cambridge. When you walk in, the first thing you notice are the furnaces. There's a large one with a door that slides open to reveal a reservoir of yellow-hot molten glass, and there are several "glory holes," the smaller furnaces used to reheat the glass on the end of a pipe or rod while it's being shaped (the name caught on because of the way the circular openings would glow brightly in dim, smoky workshops). The bank of annealing ovens against the wall, in which the freshly formed glass pieces are slowly cooled down to room temperature, look something like industrial refrigerators but have a working temperature of over 900 degrees Fahrenheit. Not for nothing is the space called a "hot shop." Being in a glassblowing studio makes it obvious just how much energy can go into creating objects.

The infrastructural networks that touch my daily life—electricity, natural gas, transportation, computation and telecommunications, water and sewage treatment—all require energy to function. Natural gas and electricity networks provide energy to be consumed at the point of use, in the form of heat and the power for appliances and devices. But the other important way in which we use energy is in the form of physical objects, all of which require energy to be manufactured and then transported to their destination. The panes of glass in the window beside me don't use any energy as they let the sunshine in, but glassblowing has made me aware of just how much energy went into creating them. There's the energy needed to power the furnaces that heat the glass so it can be shaped into useful objects, and upstream of that is all the energy that was needed to mine the sand and other minerals and process them into the raw glass. At the other end of the manufacturing process, every object has its share of the energy to move the finished goods through transportation and distribution networks to where they're used. Glassmaking might use energy in an especially glamorous way, but a similar calculus applies to the fabrication of *every* object, as flimsy and disposable as a tissue or as immovable and permanent as a highway interchange.

My colleague Benjamin Linder works in sustainable design, and he introduced me to this idea of embodied energy by running a workshop for students in my undergraduate materials engineering course. Ben brought in a variety of single-serving drinks sold in packages made of different materials, including Coca-Cola in curvy plastic, Red Bull in a slender aluminum can, and Perrier mineral water in its pear-shaped green glass bottle. The students carried out a simplified life cycle analysis for the drinks by calculating the environmental impact for each beverage container. It incorporated a number of factors, including the impact of obtaining the raw materials, shaping them into a bottle or can, and, once filled, shipping them to campus. (One of the important things that our students learn from the workshop is that understanding and quantifying the full impact of a product or process gets more challenging as more elements are considered, and that accounting for the holistic impact is likely impossible.)

Going through this exercise made me really think about what it took to fabricate the containers themselves. For example, the process of refining the aluminum for the Red Bull can—separating the metal atoms from the oxygen atoms to which they're tightly bound—requires electricity in such enormous amounts that researchers who work in sustainable design describe the material as "solid energy." The ore is shipped from wherever it's mined to aluminum smelters which, like data centers, are typically located to take advantage of abundant and inexpensive electricity. This is also why recycled aluminum has only a small fraction of the embodied energy of first-use aluminum—the energy input required to refine the metal in the first place is so high that the process of forming it into new shapes requires a proportionally much smaller amount of energy.

When we ran the analysis, I came to understand that the environmental impact of manufacturing the containers was only one element of the total impact of each single-serving drink. The label on the glass bottle of Perrier proudly notes that the water is bottled near the source in France, which means that the filled container crossed several thousand miles of ocean on a container ship before being put on a truck and driven to the store where Ben bought it to bring to our lab. The transportation of the heavy beverage, nearly all of which was powered by the combustion of diesel, turned out to be the biggest contributor to its total environmental impact. (Realizing this permanently tipped my drinking habits in favor of sparkling water bottled closer to where I live!) Infrastructural systems make it easier for us to consume energy in this subtle way, as embodied energy over the entire life cycle of a product: in the extraction and distribution of the raw materials, in the manufacturing and distribution of the product, and in its disposal— whether landfilled, recycled, or upcycled—at the end of its useful lifetime.

Infrastructure Is Power

As a many-armed infrastructure goddess, then, my hands reach out and touch a number of different systems: electricity, water and wastewater, fuel, transportation, telecommunications, supply chains for goods, and systems for the removal of solid waste. I described myself as powerful, but it's not in the sense of the magical, much less the divine. It's in the extremely techni- cal engineering sense of using power, a *lot* of power, defined as energy con- sumption per unit time.

Access to grid electricity, in particular, is remarkably close to having power in that most abstract, scientific sense. While combustion has long been the primary energy source for infrastructural systems, electricity has increasingly become the way in which we distribute and use that energy. Transmission lines make it possible to use electricity hundreds or thousands of miles away from where it's generated, and once electrons are moving through the electrical grid, they are perfectly fungible. To the end user, it doesn't matter if the power came from a coal-fired generating station, a nu- clear plant, a hydroelectric dam, or solar panels on the roof of a house. That flexibility makes it possible to evolve the energy mixture on the grid over time, as by phasing out fossil fuels and bringing in new sources, including renewables. But it's the specific characteristics of electricity, especially after two centuries of unrelenting innovation and development, that make it so astonishingly versatile. Electricity can be used at virtually any scale, from the microscopic to the huge, and it can be converted into many different kinds of energy to be used in different ways. That means that it can be used to power everything from the raw mechanical forces of giant container cranes to the tiniest glimmers of light from microscopic diodes, to produce both blazing heat and frigid cold, the sound effects in children's toys, and the sensors in life-saving medical equipment. It's revealing that we say things like "the power went out." Access to adequate, reliable electricity is having unrestricted power on tap.

Even if we're not using it in as direct a form as electricity, our power consumption is largely mediated by infrastructural systems because they're the main way in which we access a share of the energy generated and harnessed for human use. All physical systems are in some way or another about energy—how it's sourced, transformed, distributed, and consumed. The "developed" world, in contrast with the "developing world," are the places where technological systems powered by energy sources have largely replaced (and often vastly augmented) human labor.

Like all living creatures, people need energy to live, and we get that energy from the chemical energy stored in the food we eat. While it's a messy, imperfect conversion, we can estimate our power consumption from daily caloric intake, since calories are a unit of energy and days are a unit of time, and it works out to roughly the same as a modern television, 100 watts or so. Most of that energy goes to power all the biological processes that keep us alive and healthy, and some of it is used to move our bodies around. That power consumption also defines our maximum output, because we can't put more energy back into the world than we can get from our food except over short periods—this is why you need to eat more if you exercise regularly or intensely. So we can think of that baseline human energy usage as about the equivalent of having a television on for twenty-four hours a day.

But of course, most residents of industrially developed countries can look around and see that we use a lot more than a TV's worth of power every day. We use power inside our homes, in our schools and workplaces, as a share of all the energy used in the different spaces we pass through, as the power needed to move between them, and more. It's this largely unrestricted personal access to energy—whether directly or via various systems—that frees us from spending the bulk of our time on providing for the material survival needs of ourselves and the people around us. It defines what it means to live in the modern world, where it's possible to do routinely what would have been unthinkable a century ago or even just a generation ago.

Today, the per capita power usage of U.S. residents is somewhere around 9 kW (a bit over 12 horsepower)—about a hundred times the power consumption, and hence output, of our bodies alone. It's among the highest in the world, comparable to Canada, Australia, Norway, and a handful of other countries. Some of that energy, like the embodied energy in manufactured objects, we might never directly observe. Some of it we only ever use collectively, like the electricity that powers telecommunication networks or mass transit. An astounding amount of it is under our direct control, in our homes or our private vehicles. And all of the energy that's needed to make infrastructural systems function has to come from *somewhere*.

In the U.S., as for the world as a whole, about 80 percent of that energy is provided by the combustion of fossil fuels. Over the past two centuries, human societies have built out global systems to extract, distribute, and consume energy (and materials) on a fully industrial scale. Like all technologies, these systems incorporate the values of their builders. Many infrastructural systems, like municipal water supplies, include ideals like serving the public good and universal provision, or meeting the basic needs of everyone in a community. But some of the values are far less laudable: the social and environmental costs of high-quality infrastructural systems for one group are often borne by others who are not in that group. Above all, the quality of life that's made possible by a high per capita energy footprint comes with a profound disconnect between those who benefit the most and those who bear the costs. The short-term geopolitical implications of this encompass the violence and suffering associated with protecting access to coal, oil, and gas supplies. But we've also known for a hundred years that the combustion products of fossil fuel contribute to anthropogenic climate change, with everything that means for land use, food production, and more. And while the effects of greenhouse gases are global—we share the atmosphere and oceans after all—neither the level of energy consumption that led to them nor the impacts that will result are uniformly distributed. The per capita power consumption of Bangladesh, for example, is among the lowest in the world at 0.3 kilowatts, about one thirtieth of the U.S. As a low-lying coastal area in the tropics, the country is among the most at risk from sea level rise, typhoons, and flooding. And because of the close relationship between wealth, energy usage, and infrastructural systems, countries like Bangladesh have the fewest resources to mitigate or respond to the destructive consequences of climate change caused by the use of energy that they didn't even benefit from.

Energy usage has come at tremendous cost, with impacts that are still unfolding. But at the same time, infrastructural systems have provided access to energy and so enabled historically unprecedented levels of agency to more people, especially women. What might our world look like if, no matter where you happen to have been born, you had enough energy at your disposal to meet basic needs—light, fuel for cooking, energy to warm or cool your home, clean water—as well as for transportation and communications? What kinds of infrastructural systems—not just environmentally sustainable, but resilient and globally equitable—would make this world possible?

Energy is the currency of the material world. The cost of doing, making, and moving might be paid for with money, but it's priced in energy, and energy has historically been scarce and expensive. Infrastructural systems have been built to reduce the amount of energy needed to accomplish a goal (like getting clean drinking water to everyone in a city), to reduce the cost of that energy (like building pipelines to deliver oil or natural gas), or because access to energy made new ways of using it possible (like telecommunications). In other words, infrastructure often effectively makes energy more abundant, or cheaper, or both. Networks are a powerful tool for these kinds of efficiencies, which is one reason why they are so closely associated with infrastructure and, conversely, a big part of why infrastructural systems are intrinsically collective. Before we can answer the two questions posed above, then, we need to dive more deeply into understanding how and why infrastructural networks function the way they do.

Three

LIVING IN
THE NETWORKS

The flounder is a funny-looking fish.

Flounder look like they've been squashed flat, and in a way, they have been—like many of their fellow flatfish, they're born with the familiar symmetrical fish shape. But as they grow, they change shape in preparation for their bottom-dwelling lifestyle. They become lopsided. Their bodies flatten out. One side pales, and speckles appear on the other to match their sandy seafloor home. The eye that would otherwise be on the underside migrates to the upper side, leaving the fish noticeably cockeyed. All of which means that a flounder is instantly recognizable, even when it's cast into steel on a storm drain cover at the curb of a street in downtown Boston, alongside the words:

DON'T DUMP
DRAINS TO
BOSTON HARBOR

For much of the twentieth century, Boston Harbor (and the Charles River that drains into it) was famously noisome, as immortalized in the Standells' 1965 hit "Dirty Water," which Boston sports fans still sing when their team wins. Fifty years later though, the water is surprisingly clean—clean enough for abundant wildlife, including night herons in the river and seals in the harbor. Swimming in the Charles is still not a great idea, but when I was learning to kayak in Boston Harbor a few years ago, I mostly just had to worry about getting upright again every time I capsized and not about what a mouthful of seawater might do to me.

Sea kayaking in the crowded and busy harbor means learning to use landmarks to navigate, and one of the most recognizable sits across a narrow channel from the flat expanse of Boston Logan International Airport, looking for all the world like giant eggs in a carton. This is the Deer Island Wastewater Treatment Plant, and it's an important facility to those flounder, who have breeding grounds just offshore and who were in dire straits for a long time. The plant opened in 1968 to treat sewage from the whole region, and then was upgraded a number of times to meet increasingly higher standards for the wastewater it discharged to the harbor. Over the decades, the flounder have made a comeback.

Why do flounder have such a place in my heart? Besides being delightfully weird in their own right, the flounder on storm drains are a whimsical but effective reminder that we're embedded in infrastructural networks. Standing in downtown Boston, surrounded by skyscrapers and office workers, the very specificity of the fish tells us that we're part of a larger system. It's an arrow of causality, telling us that if we dump paint down the drain *here*, it will end up with the flounder (and the sea kayakers) *there*. Water, and sewage, flows downhill.

Mathematicians describe networks, in their simplest and most abstract forms, as a set of connections between nodes. Infrastructural networks bring us what we need to survive and thrive: water, of course, and energy in the form of electricity and fuel. They take away what's unsafe or unpleasant, like sewage. And they connect us to each other, through transportation and communication, linking people and places together into a single whole. The completion of a transcontinental rail link, thousands of miles long, is one of the founding myths of both the United States and Canada, and the telegraph wires strung alongside the tracks provided the earliest near-instantaneous communication across each country.

The physical connections that make up many infrastructural networks— the pipelines and cables, the sewers and ducts—are largely invisible, especially in urban areas, where they're often buried. Even when elements of the networks are in plain sight, they have a combination of ubiquity and banality that makes it easy to not really see them for what they are. The only mobile phone towers that I reliably notice are the ones that are disguised as trees. But while we mostly don't see the networks themselves, what we can easily see are the nodes, the points where we tap into them. When I'm standing in my kitchen, the faucet and drain are nodes on the municipal water network and on the sanitary sewage system, respectively; the nodes on the electricity grid are in the form of power outlets and light sockets, the blue flame on my stove is my node on the network of utility gas pipelines, and my phone jack and cable outlet are nodes on a global telecommunications system.

All infrastructural networks rely on energy to function, because *everything* needs energy to function, but not all of them use energy in the same way. Understanding networks—how they function, what they have in common and what's different between them, how they relate to energy—is a key part of understanding why infrastructural systems were built out the way they were and why they work the way they do. The networks are more than the nodes, and to start seeing them, and seeing them as a whole, I had to get out of my kitchen.

The Power of Flow

Over the course of the last few years, I've traced out most of my local municipal water network from end to end. For the water I drink at home, it's only about ten miles from the source to my tap, and then a few miles farther from the drain to the ocean—a scale that makes it pretty straightforward to grasp the system as a whole. I've taken my students to visit the town pump that provides water to our suburban college and gone to one of the regular open nights at my local water treatment plant at Fresh Pond in Cambridge. My friend Matt and I spent a day mountain biking along an aqueduct that was built to bring water into Boston from the reservoirs west of the city. My local storm drains don't all have flounder on them, it turns out, just the ones that drain to Boston Harbor. Others are labeled as draining to local rivers, like the Charles or the Neponset, and they show *different* fish, each a local freshwater species that make their home in that specific river. And, after seeing it from the harbor and from the air all those times, I went to visit the Deer Island Wastewater Treatment plant.

When I arrived at the facility, at the very start of our tour, our guide handed around a small, sealed glass jar. Inside it was a sample of the effluent that was flowing beneath our feet and into the plant. I had kind of expected sewage to be runny brown sludge, for obvious reasons, but at first glance, it just looked like plain water. I had to look carefully to see a faint yellowish tinge. Deer Island—technically no longer an island, since it was expanded with landfill and connected to the mainland—is almost at sea level, the lowest point around. If we dilute the waste enough (as we do each time we flush a toilet), the sewage will flow steadily downhill to the plant. A few well-placed pumping stations help it along, but gravity does most of the work.

To find the source of my drinking water, then, I head inland and uphill. Cambridge sends its sewage to the regional facility, but the city has its own drinking water supply. The reservoir is about seven miles northwest, visible from Interstate 95, the highway that circles around metropolitan Boston. I always notice when I'm going past because it's where signs warn winter drivers that it's a low-salt stretch of highway; minimizing the salt in the runoff that goes into the reservoir keeps it out of my drinking water. Cambridge's water supply is also part of a redundant system: it's connected (via an underground valve near my home) to the Massachusetts Water Resources

Authority network, the larger water system that serves Boston proper and many of the surrounding communities (and which also manages the downstream sewage side, including the Deer Island plant). The water for the regional system comes from two reservoirs, one about forty miles due west of the city and a larger one about twice as far west again, in the Berkshire Mountains. Whenever possible, water sources tend to be uphill of cities in order to make the most of gravity. San Francisco's water comes from the Hetch Hetchy Reservoir, high in the Sierra Nevada Mountains, inside Yosemite National Park. The billion gallons that New York City uses every day come from a dozen or so reservoirs upstate, mostly west of the Hudson River, in the Catskills.

On a muggy May afternoon, I went to visit the oldest of New York's reservoirs, in Croton-on-Hudson, just twenty or so miles upriver of the city. It was the Sunday of a holiday weekend, and the long, narrow picnic area of Croton Gorge Park was full of family groups enjoying the day. As I walked past them in turn, I heard chatter in Spanish, Hindi, and other languages that I didn't immediately recognize, and I couldn't resist looking at the tables to see the spreads of food from different cultures. The "back" of the park is a massive wall, the New Croton Dam itself. Built of buttressed blocks of stone, it looks for all the world like the ramparts of a castle except that instead of a moat, water cascades down the spillway at one end, where the regular curving terraces of the structure gradually merge into the jumbled rocks of the river. I followed a steep, narrow path to the top of the wall, where its true identity as a dam became clear. In front of me, a narrow stream of water danced and played on its way down the picturesque waterfall. To my left, the wall of the dam dropped precipitously to where, far below, tiny figures continued to enjoy their outings. To my right, its surface just a dozen or so feet below where I stood, was a small lake. It was conspicuously serene—no motorboats, no kayaks, not even swimmers. I looked back and forth from the water next to me to the ground hundreds of feet down and started to feel a bit dizzy. My vertigo was compounded by the realization that the dam, however solid and sturdy, was all that kept the placid lake next to me from becoming a crushing tsunami and overwhelming the happy picnicking families below. I shook myself a little, and reminded myself that the dam had already stood for a century and a half. Also, since it's still part of the New York City water supply, it would be regularly inspected and maintained. The undisturbed water was no accident, of course. Even though the reservoir now provides only a tiny fraction of the city's needs, access is prohibited.

My childhood was full of family picnics that closely resembled the ones I saw that day in Croton, and once, probably while my father was barbecuing tandoori chicken and my mother was managing the rest of our meal, I decided to see what would happen if I blocked up a nearby small stream. I took off my shoes and socks and stood ankle-deep in the water, happily selecting rocks from the streambed and lining them up across the current. Then I looked up, and was startled to see that the water upstream of my little "dam" was already knee-deep, getting higher and wider right in front of my eyes. I hastily pushed the biggest of the rocks away before the flooding got any worse. That was my earliest real lesson in hydrology: not only does water flow downhill, but you really can't *stop* it from flowing downhill. The best you can usually manage is to change the direction. But the flip side of this is that's also all you have to do to get the water to go where you want, and it doesn't take much. It's like a judo move for energy—you can redirect its own power.

Along with drainage, the intentional channeling of water is therefore among the earliest of human activities that we would recognize as infrastructure, whether it's irrigation along the Nile, aqueducts in Rome, or the cascading network of storage tanks across South India. Properly constructed and maintained aqueducts don't need any additional energy to function. Before AD 1300, the Hohokam people built an extensive irrigation system in which networks of canals diverted river water for agricultural use over an estimated 110,000 acres and supported the largest pre-Columbian population in the desert Southwest, in the area that's now the cities of Phoenix, Tempe, and Mesa in Arizona. Three millennia later, 97 percent of New York City's water flow is powered by gravity as it heads downriver from the Catskills all the way to the mouth of the Hudson. Once the reservoirs and aqueducts were designed, built, and opened, water from upriver began to deliver *itself* to the metropolis. To an engineer and a scientist, this is such a satisfyingly elegant solution! Whether as small as a single-serving bottle or as big as a tanker truck, delivering water in discrete quantities is expensive and inconvenient. It requires so much more energy than simply leveraging landscapes, gravity, and the way that water will flow merrily along whether it's in rivers or in pipes.

The flow of water through a network can move more than the water itself, and that energy—hydropower—can be turned into useful work. The heart of the city of Lowell, about twenty-five miles north of Boston, sits inside the triangle formed by a sharp bend in the Merrimack River and a canal that cuts across it. It was planned and built as an industrial center and is crisscrossed by canals lined with textile mills. Today, parts of that downtown core are a national historic site, and I took the train up one morning to walk along the canals and visit some of the old mills, now museums. One of the mill floors is set up as it would have been during Lowell's heyday in the 1850s, when it was the largest industrial complex in the U.S. Raw cotton, grown by enslaved people, was sent north and transformed into fabric (including "Lowell," the name for a coarse weave that was shipped back south to clothe those same people). The demonstration floor was loud, loud enough that I was glad I had picked up a pair of foam earplugs from the dispenser at the entrance. Between the noise and the long, closely packed rows of equipment, it took me a moment to realize that only a small fraction of the machines were in motion. In the middle of the nineteenth century, the operators were mill girls, young women drawn in from the surrounding countryside by the opportunities of independent employment. The power for the factory floors full of looms, carders for raw cotton, and the other machinery necessary to manufacture textiles at scale came from water-wheels spinning in the current of the canals, with the motion transmitted and distributed to each apparatus by complicated systems of shafts and belts.

The river flows and the waterwheels spin 24 hours a day, 365 days a year. The water could be channeled to where it would be used but the mills, and the entire town of Lowell for that matter, were built exactly where they were as a means of bringing together the energy of the flowing water and the machinery to use it. In the early nineteenth century, that energy could only be harnessed and moved around the mill via a physical connection, those shafts and belts. Turning a machine "off" meant detaching it from the motive source. If municipal water supplies don't require much energy to deliver water once they're built out, this went one step further: the flowing water delivered energy right to the factory and then to the factory floor. What's more, it was free. And then, at the end of that century it became feasible to convert hydropower into a form of energy that flows even more readily through a network: a spinning turbine connected to a generator produces electricity.

Water can divide and subdivide itself as it flows, distributing itself over a region. We see this in natural watersheds, where tiny rivulets and majestic rivers all find their own paths downhill, diverging and converging as they go. Municipal water and sewage networks are an artificial watershed overlaid on top of—and under—the city. Water mains and pipes subdivide the flow of clean water to buildings, and sanitary sewers carry wastewater away, often combining the flow with the runoff from the gutters, drains, and stormwater sewers. But electricity flows so well that it moves readily through transmission networks, power lines, and house wiring. It can also be divided and subdivided on much smaller scales, delivering power to the point of use inside buildings, inside machinery, and even inside computer chips. Every electrical circuit, whether visible from space or only under a microscope, is a network.

This gets us to the first reason why networks are so important when we think about infrastructural systems. If it can flow, whether poorly like crude oil, incredibly well like electrons, or even intangibly like data, a well-designed network is an efficient way to move it around, almost always requiring less energy than delivering the resource in discrete units, whether those are batteries charged with electricity or carboys full of water. Once you've gone to the (not inconsiderable!) trouble of building the pipes, getting ten quarts of water is just about as easy as getting one quart, and even having ten buildings connected to the network isn't much different from connecting just one.

That issue of flow also differentiates between concrete infrastructural networks, like water, the electrical grid, and telecommunications, and distribution networks in the form of supply chains, like the ones that bring food to the city. Fresh produce grown at a farm would typically be trucked to a packing plant to be sorted and packed and then sent on to distribution centers to be packaged into discrete units, whether punnets or pallets. These are then transported by truck, rail, or even air to a local distribution hub—the New England Produce Center is about five miles from where I live, across the Mystic River in Chelsea, in an area of Boston that's dominated by transportation links and logistics—and then out to local supermarkets. If a store puts in an order for ten boxes of cantaloupes instead of one, it will take more space in a truck, more fuel, and more labor to get them to the produce aisle. There are economies of scale to be had, and certainly having the well-defined pathway of a supply chain makes everything simpler and somewhat more efficient, but the energy costs of transportation still correspond closely to how much gets moved around and how far it goes. Cantaloupes might roll, but they don't *flow*.

Networks, Density, and the Power of the Collective

By the early nineteenth century, New York's demand for water was outgrowing the local reservoirs and wells, and the city began to build a public system to bring in clean drinking water. I think of municipal water as the quintessential infrastructural system: it needs a large upfront investment of resources to build out a transmission and distribution network, which then drastically reduces the amount of energy required to deliver water and so reduces the cost of provision. It's a human survival need, so every household on the network benefits from easy access to water but also, because waterborne diseases like cholera are contagious, from the *universal* provision of clean water. In late nineteenth-century Boston, this public health argument—that their own families would be safer and healthier because fewer people around them would be sick—was what convinced the wealthier residents to use their taxes to pay for a municipal water system and, new for the time, a sewage system to take the polluted wastewater away. It's also when the growing city began requiring all new housing developments to have water and sewage lines. Today, it's why improved water sources and sanitation facilities, including piped water on the premises, are considered to be the most basic of municipal services worldwide. Informal settlements in cities are defined, in part, by not having full access to these services, with the term "slums" reserved for the most deprived and excluded households.

There's nothing magical or inevitable about municipal water supplies, but the logic is clear: when people live in proximity to each other, the most energy-efficient thing we can do, and thus the lowest cost for everyone, is to source water in quantity and to build a network of pipes to channel it as it runs downhill, with offshoots that deliver it where it's needed. We can also build its mirror image, like veins to arteries: pipes that collect the wastewater from every household and channel the sewage downhill to where it's discharged. As drinking water and pollution standards were raised, centralized water and wastewater treatment plants were brought online. The significant—often immense—initial commitment of resources required by these systems implies two things. First, that the community that is building them has those resources and the capacity to work together to marshal and

apply them. And second, that it's willing to make an investment in the present in order to better meet the needs of the community into the future. Not
for nothing is "indoor plumbing" a metonym for civilization.

The same thing that makes it a challenge to get clean drinking water in
a city—lots of people living close together—transforms into a virtue once
you have a network. Bigger networks—that is, ones with more nodes—
amortize the cost of building out the network and any centralized facilities
over a greater number of users (which was likely part of the motivation for
building a single giant wastewater treatment plant for the entire Boston
area). And the denser the settlement, the closer the nodes are to each other,
making it easier to connect them up. Together, these two features are more
than just economy of scale. They characterize an economy of networks, and
help explain why infrastructure is so closely associated with cities. When
population centers like New York City began to bring in water from outlying
reservoirs, those aqueducts were the beginning of a network and those economies began to kick in.

Conversely, however, if the potential users of a service are far away from
each other, it becomes more challenging and expensive to connect them.
Solving this "last-mile problem" was part of the impetus for the Rural Electrification Act of 1936. Franklin Roosevelt's New Deal included a program
for nationwide electrification by building out dams, power plants, and transmission lines, which also provided employment to workers left jobless by
the Great Depression. Federal loans for local power distribution were made
available, but widely spaced farms weren't seen as a viable market by the
mostly city-based utility companies. Instead, farmers banded together to create local electricity cooperatives. By 1959, 90 percent of farms were connected, up from just a few percent at the start of the program. To this day,
most rural electricity in the U.S. is supplied by not-for-profit co-ops that got
off the ground with this federal funding. In 1949, the act was extended to
cover phone lines and then, half a century later (2008 and 2014) provisions for

broadband Internet access for rural areas were added. The last-mile problem means that these regions were (and many still remain) underserved by investor-owned providers.

Flow, size, and density all contribute to making networks a highly effective way to move resources around, from both an energy and an economic perspective. But the connections between nodes can matter too—where they go, how far, and what happens when they're disrupted.

Bottlenecks and Interconnections

Across the ocean from my home, on a glorious spring day, I took the train north out of London to the town of Tring, in the Chiltern Hills, to walk in the footsteps of our Bronze Age ancestors.

Roads are perhaps the oldest systems that we would recognize as infrastructure. Humans have created paths for their entire existence as a species, just as other animals do, and for the same reason: as anyone who's ever gone for a hike appreciates, it takes less energy to follow an existing trail than it does to create a new one. The act of following it often makes the trail wider or more distinct, and easier still to follow—to borrow the words of Paulo Freire, we make the road by walking. That morning I was on the Ridgeway, a trackway that humans have been using for five thousand years. The fields and villages of southern England stretched out below me, a patterned green carpet under a blue sky. Besides yielding breathtaking views of the countryside, the high ground made it easier to travel without getting bogged down in mud. On the western side of the Atlantic Ocean, the Natchez Trace traverses Tennessee, Alabama, and Mississippi and is among the oldest surviving routes in North America, estimated to date back ten millennia. Roads are a staple of postapocalyptic futures, too. Even in fictional settings where civilization has collapsed and cars rust in heaps, networks of roadways still provide routes across the landscape.

If we think about infrastructural systems as moving resources to where we want to use them, then they are inextricably about location on the landscape, specific points on the planet. At a minimum, those points are *us*—for all that we talk about being online and in the digital sphere, each of us has a body that is located in a particular place in the world. In other words, we are all nodes. And even the simplest connections between nodes forms a network. That's the kind of connectivity that roads enable.

In the United States, one of the largest countries in the world by land area, it's almost hard to appreciate the value of these connections. It's not just that good roads and highways reduce the time and cost required to move goods and people around. At an even more basic level, it's mostly taken for granted that almost any point in the country is on the road network. When I moved to Seattle for a year, I didn't think about *whether* my new apartment, three thousand miles away, was reachable by car from the street that I could see out my window in Cambridge. I just had to figure out what route to take. In network science terms, that means that the American road network is mostly connected. There are a few exceptions, of course, like islands. When I left the Honolulu airport after the long flight over the Pacific from Los Angeles, I couldn't help but laugh at road signs that were the familiar blue-and-red shields of an *interstate* highway. It's when I first learned that these highways are defined by federal funding and adherence to a set of design standards, not whether they cross a state line, and also that they're officially "interstate and defense highways"—which might help explain their presence in Hawaii, home to military bases, including Pearl Harbor.

The less obvious exceptions to the general connectedness are "practical exclaves," places that aren't directly reachable from the rest of the country. Alaska is a practical exclave, but so is the tiny town of Point Roberts in the state of Washington—like Alaska, it's only accessible by sea or by going through Canada, and it's long been rumored to be a popular place for Americans in the Federal Witness Protection Program to settle because getting to it by car from elsewhere in the U.S. means going through two border crossings.

Even if it's not on an island or in an exclave, each point in the network of roads is not as well connected as every other. East of Seattle, Interstate 90 climbs up through the Cascade Range at Snoqualmie Pass. As one of only two interstate highways into the region, it's a critical link, carrying nearly thirty thousand vehicles every day. I've lived in places with snowy winters for nearly my entire life. But it was only during that year in Seattle that I learned how to put chains on my tires. If I wanted to go snowboarding when it was snowing, I'd have to go through a checkpoint where state police officers confirmed that every vehicle headed toward the pass had good traction. My familiar Boston end of I-90 is embedded in a network of surface roads, so traffic can be routed around accidents on the highway. At the Seattle end, where the mountains limit the city's connections, anything that obstructs the highway is a much bigger logistical headache. It made sense to ensure that drivers headed up from the rainy coast were prepared for the snowy conditions.

Headed north from Seattle, the U.S. road network merges seamlessly with Canadian roads at the border. I've taken the train between Seattle and Vancouver, British Columbia. It's a stunningly gorgeous route, with high mountain peaks on one side and the Pacific Ocean on the other. The western rail lines were built with the same kind of tracks (standard gauge, with rails 4 feet, 8½ inches apart), so railcars have always been able to move freely between the two countries—for all that it crosses a national border, it's a single network. In contrast, before the U.S. Civil War, trains in the American South mostly ran on a broad gauge, with rails five feet apart. Some cities like Montgomery, Alabama, were served by two different rail companies, with separate stations on different gauges. That meant that transshipping goods between the two networks involved the time-consuming and expensive process of unloading everything from the first train, transporting it across town in carts, and then reloading it into railcars on the other line. On top of this, many of the rail lines in the South were isolated railway spurs dedicated to taking cotton from the fields to the ports, rather than the much more connected network of the North. This lack of connectivity had a severe impact on the ability of the Confederate States to move soldiers and supplies during the Civil War. After the war ended, trade between the North and South increased and the hassle and cost of the break of gauge became increasingly untenable. In 1886, after months of planning and preparation, all of the broad gauge southern track was changed to standard gauge in just thirty-six hours: tens of thousands of workers pulled up one rail of the track, moved it a few inches closer to the other, and spiked it into its new position. Railcars got new undercarriages, and the two separate smaller networks became a single, fully interconnected, national network.

When the networks that allow people and goods to move around easily are disrupted, as by a natural disaster, this process is reversed and communities can find themselves as tiny, isolated networks. In September 2017, the main bridge for the rural community of Rio Abajo in the Utuado region of central Puerto Rico was washed out by Hurricane Maria. In the immediate aftermath of the storm, residents ran a line across the river, and suspended an open wooden box from it by pulleys, just large enough for one person. It was the most tenuous thread of connection to the networks, systems, and supply chains of the larger world. Without power, young locals took turns pulling the box across, hand over hand, to carry residents across the river to the main road, and to bring food and other basic supplies to the residents of Rio Abajo. Including, of course, cases of bottled water.

Networks and Directionality

Infrastructural networks are effective ways to move resources around because they allow for flow (and the energy efficiencies that result), and they benefit from density. But the directionality of the flow in networks also affects the kinds of value that it can provide to each user.

The gas stove in my kitchen is the old-fashioned kind, which means that every once in a while, I come home to the faint but unmistakable rotten-egg smell of mercaptan, the chemical compound that's added to natural gas for just this purpose: it makes even tiny gas leaks, as when a pilot light has blown out, detectable. (Large, dangerous leaks are unignorable, the olfactory equivalent of a blasting alarm.) Utility gas goes on a one-way trip. It's carried through gas mains to local lines to the building I live in, then to my stove, and then to the burner where it meets its fiery end under a saucepan. The molecules of gas are then transformed, producing the heat that I need to cook my food in the process of being converted into (mostly) carbon dioxide and water vapor and released into the air. Grid electricity works this way as well. It needs a closed circuit for current, but the actual energy goes on a one-way trip, from the power station where it's generated to the point of use where it's turned into a different, useful form of energy—like light, heat, sound, mechanical movement, or something else—and dissipated.

Water doesn't evaporate when it arrives at my apartment, but it still makes a one-way trip that ends in a transformation. It might even happen in a split-second of freefall: what comes out of my faucet is potable water, but when it reaches the drain just below, it's wastewater. The water started its journey at the reservoir and arrived at my home via a network of water mains and pipes. As sanitary sewage, it continues to flow downhill, converging on ever-larger pipes until it arrives at Deer Island to be treated and returned to the outside world via diffusers that look like giant salt shakers sitting on the bottom of Boston Harbor, nine miles away from shore. The whole way, it's all traveling in one direction. Natural gas, water and sewage, cable TV, and the circulatory system for blood in our bodies—which functions in almost exactly the same way as water and sewage—are all examples of one-way networks. While one-directional networks can have many nodes that are close together, those nodes aren't connected to each other: the sewage that goes down my drain never touches other homes, at least not unless something has gone very badly wrong.

But other kinds of networks allow for connections *between* nodes, and that's what makes them useful. Back when they first came out, it only made sense to buy a fax machine if you already knew at least one other person who had one so you could send messages back and forth, if only to that person. But as the number of people with fax machines—the nodes in this network—increased, the number of connections between them increased rapidly: with two fax machines, only one connection is possible, but with three fax machines, there are three possible connections; with four machines, six. With only five machines, there are ten different connections possible; by the time you get up to ten machines, there are 45 different ways to connect them. The numerical relationship between the nodes and the ever-increasing connections between them (on the order of the square of the number of nodes) was codified as part of Metcalfe's law, named after Robert Metcalfe, one of the inventors of the Ethernet communication protocol for computers. Metcalfe argued that the value of a network to its users—whether fax machines, phones, or computers—increased proportionately with the number of connections to each node. The more possible connections there are, the more valuable it is to *each* user.

Nodes on one-way networks are like leaves on the spreading branches of a tree, but two-way connections create something more like, well, a net. When all the nodes are connected to each other, the possibilities quickly explode. Roads and other transportation networks work like this, as do communications networks, including postal mail, telephones, and the Internet. Not only do users benefit from being connected to the network themselves but also, broadly speaking, individuals benefit from having more users on the network: more people and places that they can connect with, talk to, visit, send goods to, and so on. So the more people who have access to the system, the better off everyone is individually. This is also true for potable water and sanitation, but for a different reason: because waterborne diseases are contagious, each individual person is safer if fewer people around them are sick. We're now starting to edge into why infrastructural systems are often considered public goods. The value of connectedness is a driver for networks to be as large and as widely adopted as possible, whether by making them available to more users (more nodes) or by making them interoperable and so forming a larger network (as when railways moved to a single gauge).

Scaling Up Networks

The water and sanitation systems where I live are charmingly small, almost quaint. From the reservoir where my drinking water comes from to the point where the treated wastewater is discharged into the ocean is about the same distance as the Boston Marathon. Much of the network was designed and built a century or more ago, so it's largely gravity fed, not needing much additional energy to function. It's a human-scale system, which is part of what made it so easy and fun to explore, by car and bike and subway and sea kayak. But it's also something of an anomaly. Most infrastructural systems don't lend themselves to being apprehended so easily or completely, not even ones as seemingly straightforward as water provision.

A few years ago, I spent a May weekend driving around California with my friend Charlie Loyd, an expert in satellite imaging, to explore different parts of the state water system, including a visit to the San Luis Reservoir, the fifth-largest in the state. It's in the mountains above Interstate 5 due east of Santa Cruz. We parked at the side of the adjacent road and walked down to the water's edge, listening to the low drone of wind turbines on the ridge blending with the susurration of the drought-yellowed grass all around us, intermittently enlivened by birdsong. We were standing in an interlocking system of rivers, dams, reservoirs, pumps, aqueducts, and canals that spans many states—seven in the Colorado River consortium alone—and thousands of miles, reaching south to the border with Mexico. Household usage only makes up a small part of it; the system also delivers water for agricultural irrigation, including the thirsty but enormously productive Central Valley in California, and produces hydroelectric power, as at the seventeen generating turbines at the Hoover Dam in adjacent Nevada. The water doesn't just flow downhill: the California Aqueduct, for example, goes up and over the Tehachapi Mountains, thanks to the Edmonston pumping station. Fourteen six-story-high pumps can push two million gallons of water per minute up two thousand feet at the highest single-lift plant in the world, helping to make the California State Water Project the single largest consumer of energy in the state.

The year after visiting the reservoir, some other friends and I walked the full fifty-mile length of the Los Angeles River over the course of a long weekend, from the western San Fernando Valley south to the Port of Long Beach. On our first day, we stopped at the William Mulholland Memorial Fountain and poured some Los Angeles tap water on the marker for the man who was the primary architect of the city's water supply but whose legacy is complicated, to put it mildly. As chief engineer of the city water system, he oversaw the construction of the Los Angeles and the Colorado River Aqueducts that bring water to the city from hundreds of miles away. Mulholland was also responsible for the St. Francis Dam, which failed catastrophically in March 1928, sweeping hundreds of people to their death.

We were walking the length of the river in part because it's a tiny piece of the California water system that *is* on the scale of our bodies, something that we *could* encompass. While a few stretches of the riverbed are accessible and safe, people have generally been discouraged from entering the river, and definitely from following it on foot for three days straight. The waterway is lined with concrete for most of its length and serves as a storm drain for the city, collecting rainwater that runs off many square miles of paved-over urban landscape and carrying it out to the Pacific Ocean. That sprawling urban watershed makes the channelized river susceptible to flash floods. Late in the afternoon of our second day of walking the river, it started to rain—nothing serious, just a gentle patter. We were in a stretch where the channel was narrow and the walls on either side of us were vertical, a human-made slot canyon. The water was deeper than our hiking boots, and we couldn't know if there was more coming from behind us. If there was, it could arrive fast and hard, and the steep walls meant we wouldn't be able to get out of its way.

At the San Luis Reservoir, as Charlie and I walked down to the shoreline, the ground beneath us abruptly became mostly bare rock and soil, with a few small, fast-growing plants, lost fishing lures, and patches of mineral salt. We were there in the spring of 2015, and that last fifty paces or so to the water's edge was an ongoing drought made visible. It was what the absence of nearly 40 percent of the reservoir's capacity looked like. But that water wasn't missing because of the weather conditions per se. It was missing because the reservoir is part of a much, much larger system. As networks become physically larger and technologically more complex, with greater energy demands, the social context can similarly become more and more complicated. Understanding even just the legal issues around water in California itself is a lifetime's study. While talking through what we had seen over the course of that weekend, Charlie and I landed on the term "ultrastructure" to describe this web of social structures, all of the cultural, political, regulatory, and other systems that shape and govern infrastructure. In this case, that would include where the water gets delivered, and how it's used, and by whom. Drought or not, filling the reservoir or not, what made it "missing" water was that it was part of the human system of ultrastructure, promised to someone for their use. Networks are inherently collective, not individual. They are about doing things together, which means that they are necessarily shaped by social systems. But the converse is also true: technologies develop into infrastructural systems because of the evolving roles they play in our lives.

Eight months later, my friends and I in the Los Angeles River were fortunate—we shouldered our packs, put our heads down against the rain, and kept moving at a brisk pace until the river channel again widened out into gentle sloping banks. When it finally did, we only needed to walk a few feet out of the water before we could breathe a sigh of relief that we hadn't wound up on the wrong side of a system whose behavior we only understood imperfectly. Larger, more complex systems are harder to understand, in part because they can behave in counterintuitive ways, especially when they interact with human systems and preferences. What's more, infrastructural systems aren't isolated from each other. They often depend on other infrastructure to function, at least road access and power from the electrical grid, but they can also be deeply entangled with each other. Researchers talk about the "food-water-energy nexus" because of the close relationship between the three, especially in places like California, where agricultural production is linked to the electricity grid: the same water irrigates fields, generates electricity, and consumes energy as it moves through the system.

For all of its uniqueness, water in California is also typical of many of the infrastructural networks that we are embedded in. They are huge in extent, spanning multiple jurisdictions, and incorporate elements that are physically enormous and that require significant amounts of energy to function. They're also fiendishly complex, often under continuous oversight in order to balance multiple competing needs. One story of infrastructure in the twentieth century is how these systems went from small-scale networks—conceived and overseen by architects like Mulholland and largely comprehensible in their entirety by individuals—to the ones that surround us today. Thinking of infrastructure as physical systems of networks and energy is only part of what we need to understand them as they exist today, much less the directions they'll be taking in the future. They are also human systems that incorporate particular ways of seeing, organizing, and acting in the world.

Four

COOPERATION ON
A GLOBAL SCALE

One of my very favorite sounds in the world is made by split-flap displays, also known as Solari boards after a well-known manufacturer. The "screen" shuffles to show each of the letters and numbers in a row of text, pauses to show the information, and then shuffles again—*clickety-clickety-clickety*—to reflect updates. When I was growing up, these were the standard displays for arrival and departure information at train stations and airports, but they've become rare. While bright digital screens are more flexible, they update silently, without providing that environmental cue to waiting passengers to look up and see what's changed.

Among the few remaining Solari boards that I know of is the one in the gorgeously rebuilt San Francisco Ferry Terminal on the Embarcadero, and whenever I find myself in the city, I try to go across the bay from there to Oakland. I buy my ticket and listen to the sound of the ferry times and destinations updating while I wait to board. No matter what the weather is like during the trip—cold and wet or sunny and warm or, seeing as it's San Francisco, all of the above—I stand on the deck so I can take in the Pacific Ocean, the Marin Headlands, and the beauty of the city in its setting. As the ferry steams away from shore, a neon sign catches my eye, glowing red and spelling out the words HILLS BROTHERS COFFEE on a large brick building at the water's edge. It's a ghost sign, a holdover from the city's heyday as a port, when ships loaded with sacks of beans would arrive and the coffee would be unloaded and roasted before being sent onward to the rest of the U.S.

Many similar late nineteenth- or early twentieth-century buildings located near the shoreline in San Francisco, with their high ceilings, exposed brick, and large windows, were once warehouses. The area is recognizably kin to districts found in other cities in North America, not just on the sea-coasts but also in Chicago, Toronto, and Minneapolis on the Great Lakes. Ships arrived and tied up at the docks, and stevedores would unload the break-bulk cargo, the barrels, boxes, bags, and bundles stowed in the hold to send them on further by land. In the other direction, goods arriving at the port by truck or rail were loaded onto ships for the next stage of their journey. All of this required time and coordination. Goods needed to be sorted and stored while waiting to be loaded for transshipment, which is where the warehouses came in. As shipping traffic grew during the first half of the twentieth century, the docks and warehouses became busier and busier, hives of activity twenty-four hours a day. But then, in the late 1960s and early 1970s, warehouse districts began to hollow out. In the U.S. and Canada, as elsewhere in the world, they stood empty or were informally occupied for decades, by squatters or as live/work spaces by artists. The resurgence and growth of urban centers and the attendant rise in property prices incentivized the zoning and redevelopment of warehouses for residential or commercial use. In San Francisco, that meant loft spaces for galleries and tech startups, and also tech giants—the former Hills Brothers Coffee roasting plant is now home to Google offices.

The explanation for why all of these warehouses emptied out half a century ago is more clearly apparent at the other end of that ride from San Francisco. After sailing under the Bay Bridge, the ferry enters a narrow channel lined with shipping cranes that look for all the world like giant spindly steel beasts at the water's edge, necks extended to drink. For the last ten minutes or so of the ride we glide past the docks of the Port of Oakland. I always gawk at the huge ships with their multicolored boxes.

The invention, refinement, and uptake of the intermodal shipping container—those standardized, stackable steel boxes—changed nearly everything about transporting goods. Containers could be filled and sealed at the factory and transported to the docks by truck or train, then loaded by crane onto cargo ships, stacked on the deck and locked into position for the voyage. On arrival, the same process happens in reverse: cranes lift the containers from the ship and place them directly onto truckbeds or on trains, so the cargo can continue overland to its destination. Containerization meant that the labor-intensive, time-consuming, and sometimes "leaky" (it was easy for loose cargo to go missing) process of unloading or loading ships at the docks became much faster and required fewer workers. Instead of manual labor, the energy for moving the cargo is provided by the container crane, occasionally from diesel generators but usually as shore power, from the electrical grid. Despite strong resistance by labor unions, within about twenty years almost all cargo in the world was containerized.

Intermodal containers made it possible to seamlessly connect two wildly different worlds, maritime shipping and ground transportation. The move to containerization reshaped cities as shipping traffic relocated to ports that were more compatible with these new modes and those ports grew larger in turn. The inconvenience of transporting cargo overland out of peninsular San Francisco made the mainland more attractive, which led to those vacant warehouses on the west side of the bay and container cranes in Oakland. That ease of transferring containerized goods between ships and trains or trucks—needing less time, labor, and warehouse storage—meant that maritime and overland transport networks effectively merged into a single, transmodal network. Containers didn't just lower the cost of shipping—they all but eliminated it. Transportation costs became negligible, no longer a barrier to worldwide distribution. Suddenly, it didn't really matter where a product was made, or how far away the factories were. Once the process of transferring containers from ships to ground transport was swift and reliable, long-distance supply chains suddenly became feasible. International trade agreements soon followed and global trade exploded.

Standardized shipping containers created a low-friction planetary-scale network for physical goods. This would turn out to be a key element of globalization, including the idea of "offshoring," eventually reshaping communities worldwide. We're still sorting through the consequences of this.

Standards as Enabling Global Cooperation

If we're embedded in networks that span the globe, that means there must be some way to coordinate across them so that people can cooperate on their function despite living in different places, speaking different languages, and having different cultural backgrounds. Much of this coordination happens via standards, which are carefully defined and agreed-upon ways of interacting with, well, everything, but especially technological devices and the material world.

It's a movie cliché that when someone wakes from unconsciousness, their first question is "Where am I?" and the second is "What time is it?" Global mapping and standardized time are so deeply foundational to how we think about time and space that it's easy to forget that they aren't inherent to the world. They're standards, and they came out of the growth and spread of transportation and communication networks.

The clocks at the Royal Observatory at Greenwich have served as reference standards for navigation since the late seventeenth century, when the first chronometers made it possible for sailors to calculate their longitude relative to the prime meridian. Despite this, every place in Great Britain still set its time independently. Local noon was based on the highest point of the sun in the sky, which varied by tens of minutes across the island. As cities began to be connected by rail—swifter, serving more people, and on fixed lines—the increased risks of collision meant that the time differences became dangerous as well as cumbersome. Beginning in 1840, railway companies began to run their operations on a standardized time. Initially, staff hand-carried synchronized portable chronometers on the trains to disseminate "railway time" to stations, but within fifteen years, timekeeping signals were sent out from Greenwich by telegraph. Over the course of the next half century or so, the growth of rail networks across Great Britain drove the adoption of a single time across the island. The modern system of time zones gained traction worldwide as undersea telegraph cables and then radio transmissions cemented near-instantaneous communications channels across the far-flung British Empire. By 1929, universal standard time had gone global: synchronization within each time zone, with a fixed offset based on the longitude relative to the prime meridian.

In parallel with the spread of a coordinated time system, the nineteenth century also saw the rise of mapping. Beginning in 1747, the British Army was commissioned to make its first maps, initially to aid in finding dissenters and quelling military uprisings in the Scottish Highlands. (To this day, the UK mapping agency is known as the Ordnance Survey.) Through the middle of the nineteenth century, all of Ireland, England, and Wales were surveyed and mapped, largely for taxation but including higher-resolution maps of urban areas to support the construction of municipal water and sewage systems. The commercial possibilities of maps were quickly recognized. Railways and other companies pressed the government to create more and better maps, and to make them broadly available. Across the ocean, the American Revolution ended with the Treaty of Paris in 1783. Nearly the first thing the new government did to assert control over the newly ceded territory was to commission a cadastral survey—a map that showed how the land was to be divided up into discrete parcels, a shared system in which to describe and sell it. The surveying process continued alongside U.S. territorial acquisition, first the Louisiana Purchase of 1803, followed by western expansion through the nineteenth century. The rectangular grid that the survey produced, each square a mile on a side, was used as the basis for the colonization of the western half of the continent by European-descended settlers, and it still literally shapes how the land is used today. At night, from the air, the work of the surveyors is visible as a sparse, orthogonal grid of lights. I can always tell when I'm getting close to home on a cross-country flight, because the orderly squares of the western U.S. begin to transition into the dense, irregular network of roads and streetlights of the northeast. On the other side of the planet, the Great Trigonometrical Survey of India was begun in 1802 and would also continue through most of the century. British surveyors lugged half-ton theodolites from the southernmost tip of India, Kanniyakumari (then known as Cape Comorin), all the way up the spine of the subcontinent to the Himalayas and Mount Everest, measured as part of the survey and named after an earlier Surveyor General of India. Defining this "Great Arc" was part of a scientific quest to determine the exact shape of the planet, but mapping the subcontinent was primarily a commercial and a military imperative for the East India Company and then for the British Raj. On all three continents, mapping was a means of increasing legibility—knowing what was on the ground, and where—as a way of increasing control and access to resources.

The glowing blue dot on my phone that tells me where I am today is the product of a later iteration of the same story. The Global Positioning System (GPS) relies on a constellation of satellites that can triangulate the position of a spot on the earth in order to locate it precisely. To do this, it depends equally on the exquisitely accurate geodesic model of the planet that evolved from those nineteenth-century trigonometrical surveys and on fantastically precise clocks, in satellites and their ground stations, that are the descendants of those Greenwich-synchronized railway chronometers. Developed by the U.S. military, GPS was first opened to civilian use in the late 1980s after a commercial Boeing 747, Korean Airlines Flight 007, drifted off course and into prohibited airspace over what was then the USSR, where it was misidentified as military reconnaissance and shot down with no survivors. Initially, only a degraded version of GPS was made publicly available, with "fuzzed" data that reduced the precision to a few hundred feet. But, as with high-resolution maps in England two centuries earlier, the commercial value of GPS became clear and in 2000, the full precision of less than a foot, and even down to a fraction of an inch under some circumstances, was made available to all users.

I think of these two systems, time and mapping, as the infrastructure for infrastructure. Their growth in extent and precision has paralleled and enabled the infrastructural networks that rely on them. Today, they make up a unified, precise, global system for locating points in time and space that underpins virtually all other forms of technological and social coordination.

Shipping containers are a physical instantiation of a standard: they are the right size and shape and have the right fixtures to be lifted by a crane and secured in place on a container ship, a truck, or a train. As long as the box itself has the right physical properties, including its dimensions, mass, and rigidity, the actual contents are largely (and usefully) immaterial. Standards are what make it possible for different systems to connect with each other and work together. Time and mapping are the keystones, but standards of every sort can be found everywhere, quite literally. Within arm's reach of where I'm working, I have a desk calendar with dates, clocks showing the time, and a lamp with a lightbulb, power plug, and USB outlets. My papers are a standard U.S. size, 8½ inches by 11 inches, and the sheets fit in the printer I have at home and in those I use on campus. There are hundreds of different technical standards implemented in my laptop, and I'd imagine at least as many in my smartphone. Even the butter on my toast is governed by a standard: I wondered why butter in the UK and Europe tasted better than at home in the U.S. and learned that the minimum allowable milkfat percentage is higher. The International Standards Organization, a nonprofit oversight body, has defined more than *twenty thousand* standards, which means there's almost certainly a standard that's relevant to anything anyone might want to do, including one for my cup of tea (ISO 3103).

Since standards are used to make systems work together smoothly, we mostly only notice them when there's a mismatch. When I moved from Canada to the U.S., and then from Boston to Seattle, I brought household appliances with me with no concerns about whether they'd work in their new location or, worse, that they'd be dangerous. When I moved to London, though, I left them all at home. Besides using differently shaped plugs, wall outlets in the UK are at twice the North American voltage and deliver correspondingly higher power. I'm reminded of this every time I'm visiting and put an electric kettle on, because it comes to a boil startlingly faster than I'm used to. Unlike household appliances, though, electronic devices like my computer and phone are on a global standard. It's why they usually come with swappable chunky adapters that transform the high voltage and alternating current of local wall power to the lower voltage and steady direct current that they run on.

Payment systems, including credit and debit cards, adhere to international standards so that they function in many different places as long as there's access to a telecommunications network and the electricity to power it. Cash itself is rarely interoperable across national borders: regional currencies like the euro are among the few notable exceptions. The closest thing to a global currency standard is still the U.S. dollar, so much so that hundred-dollar bills are reportedly the most counterfeited in the world. But it's the *idea* of money—that you can exchange money in some form, however concrete or abstruse, wherever in the world you are, for whatever goods or services you need or desire—that might be the most important international "standard." It's the default structure of cooperation that's used on a global scale and it's what makes it possible to operate networks that cross almost every sort of political boundary.

A unified global system lets a passenger plane land in Chicago or Lagos or Beijing and safely take off again. Standards provide a way to hide, and therefore manage, the complexity of systems. At least in principle, and mostly in practice, as long as the relevant element adheres to a mutually agreed-upon set of rules, the specifics of just *how* those rules are met doesn't have to be taken into account.

Standards are also a key element of infrastructural systems because the decision to do things the same way, or at least interoperably, is often what makes it possible to scale networks up. It's how the standardization of rail gauge made the northern and southern networks in the U.S. into a single, national railway network. Standards are why the U.S. and Canada can share a roadway network and an electric grid across state, provincial, and national borders, and standards are how shipping containers unified maritime and overland transportation networks globally.

But there's a third, more subtle way in which standards underpin infrastructural systems. Over the past century or so, infrastructural networks have grown in spatial extent, from localized systems to ones that now span the globe. The growth of these networks has been accompanied by a change in scale—nearly every element of these systems is bigger and consumes more energy—and by a corresponding increase in technological complexity. Standards make this complexity possible.

Higher and Higher:
The Trajectory of Technology

One February morning, I woke up before dawn to go watch the arrival of the ship CMA CGM *Benjamin Franklin* at the Port of Oakland. It had launched just a few months earlier, in December 2015, and was making a round of visits to deepwater ports on the U.S. west coast. The *Benjamin Franklin* is among the largest container ships in the world—at just shy of four hundred meters from bow to stern, it has a carrying capacity of 18,000 twenty-foot equivalent units (TEUs), the standard unit of measure for intermodal containers. As the sun rose, the ship's quarter-mile length came down the channel and a pair of tugboats snugged the vessel up to its berth beneath the container cranes of the port, ready to be unloaded.

What makes this ship different from the cargo ships that once tied up across the bay in San Francisco is not just the stacks of shipping containers on its deck. There's an important hint in the scale. Everything about it is vastly larger than the ships of a century before, with the exception of its human complement—the *Benjamin Franklin* carries just twenty-six crewmembers. If we're trying to understand the world of networked technological infrastructure in which we live, the networks are only part of the story. The other part is the *technology* itself—the physical systems, not just of the networks themselves but all of the technological devices and facilities necessary for their operation and use, whether they're container ships or electrical substations or cellphone towers or wastewater plants. Over the course of the twentieth century, there was a profound shift in our material technologies. In order to recognize that shift and what it means, we need to think about what technology is and how technologies evolve over time.

I enjoy knitting by hand, and my preferred needles are smooth bamboo, machine-made and imported from Japan. But I don't actually need specialized knitting needles. It would be more challenging and less pleasant, but I could still knit if all I had was two straightish sticks. Similarly, I usually knit soft, brightly colored machine-made wool, but I could instead use wool that's been carded and spun by hand and left undyed. I can imagine harvesting long grasses and braiding the fiber into "yarn," or I could go lower tech still. Because here's the thing: even if all I had was two sticks and a length of vine, pulled off a tree and stripped of leaves, I could still do something that is recognizably *knitting*, however messy and imperfect the result.

In *The Nature of Technology*, W. Brian Arthur makes the case that all technology has three elements. The first is that there must be some physical phenomenon at its root. Money, for example, is *not* a technology. It might take the form of a string of seashells, or of printed pieces of paper, or of bits inscribed in a computer database, but the actual idea and function of money is independent of any specific physical phenomenon. Legal or educational or political systems might be enabling structures, but they are *social* systems. Knitting, on the other hand, *is* a technology because it relies on particular physical phenomena. In order for me to knit, the needles and the yarn need to have specific material properties, so that I can use the stiff needles to pass the long, flexible yarn over and around itself. The knitted piece that results is a large, flat knot, with physical properties (like stretchiness) that come along with that topological structure.

The second characteristic of technologies is that they are combined and assembled out of other technologies. Instead of knitting with two sticks and a length of vine, I go to my local crafting store and choose what kind of knitting needles I want to use: steel or bamboo or plastic, small or large, straight, double-pointed or circular. I can select from many different yarns, not just wool but cotton, linen, silk, and a wide array of synthetic fibers like polyester. If I wanted to crochet instead of knit, I would recombine these technologies to create a new technology. I might use the same yarn, but I'd use a different tool: instead of straight needles, a crochet hook that catches and pulls. I'd also use a different process, which would produce a different kind of three-dimensional knot.

The third characteristic is, to my mind, the most interesting and subtle: technologies are *recursive*. If you look closely at any given technological artifact or process, you can observe that it's created out of other technologies, and the same is true if you look more carefully at *that* technology, all the way down to some basic physical phenomenon. Even something as simple as a knitting needle can incorporate an extraordinary range of technologies in its manufacture. That might include growing the bamboo, or pumping the crude oil and chemically transforming it into smooth inert plastic, or mining iron ore, refining it, and making steel, followed by the different fabrication processes used to shape and finish the needles. We can tell similar stories about the yarn: different fibers might incorporate different processes. For wool, they include raising and shearing sheep, carding and spinning the wool before coloring it in a range of bright hues with synthetic dyes, then gathering and packaging the yarn into the soft skein I pick out at the store. And if we look closely at any of these component technologies—from sheep husbandry to dye chemistry, even the hand-powered winder I use at the store to turn the skein into a compact ball of wool—we can see that each subtechnology is itself composed of further subtechnologies.

This becomes even more obvious when we go *higher* tech. Industrially manufactured knitted clothing is ubiquitous; I doubt there's been a single day in the last year, maybe the last decade, that I didn't wear something machine-knit, like a T-shirt or socks or leggings. A few years ago, I had the opportunity to visit a local textile mill, and even the relatively simple industrial knitting machines on their factory floor were the size of a car, cost thousands of dollars, and required skilled operators and specialized maintenance. But at their root they still relied on the same basic physical phenomena as hand knitting, of needles entangling yarn with itself into a continuous fabric.

Once we think of technology in this way, as combined and recursively assembled out of existing technologies, we can see why we can usually tell if a given piece of technology is older or newer. Technological development has a direction, because complexity tends to increase, not decrease. Elements can be combined in new and different ways: a shaft and gears can connect a waterwheel in a river to a roomful of looms in a textile factory instead of grindstones in a flour mill. A familiar physical principle can be used in a new way: even though it's been known since antiquity that steam could push on something to make it move, it wasn't until the steam engine that this behavior was harnessed to spin a turbine. A newly discovered physical phenomena can be harnessed, to new ends. In 1831, at the Royal Institution of Great Britain, Michael Faraday demonstrated that moving a magnet in a coil of wire creates an electrical current. He went on to use this principle to build a dynamo, the earliest electrical generator. Put new discoveries together with the old, and a new technology emerges. Burning coal to produce steam to turn a turbine that moves inside a generator makes a rudimentary thermal power station. Or attach the generator directly to a waterwheel, spinning in the current—hydroelectricity.

Newly discovered physical phenomena can also lead to completely novel technologies. Solar panels are made of materials that produce electricity when they're exposed to light, an entirely different process from Faraday's electrical generators. But new technologies also need to build on what's already in use. The photoelectric effect was first observed and described in 1839, and the first rooftop solar panels in New York City were installed in 1883, but the widespread adoption of solar panels in their current familiar form had to wait decades until all the associated technologies, like advanced methods for refining high-purity silicon, were developed. This demonstrates that the particular set of technologies available at any moment is *contingent*— a different set of technologies in the past, or different scientific discoveries, or even the same discoveries in a different order, would mean different technologies in use. It's the premise of a lot of alternate history science fiction. At the same time, new technologies are built to interface with existing systems, especially infrastructural systems—to run on AC wall power, for example— which also shapes how they develop. There's contingency but there's also *continuity*. This path dependence means the infrastructural systems that are in use shape the technologies that are developed, and vice versa.

I own a handknit wool shawl that I bought while visiting the Faroe Islands, made of undyed wool from the hardy local sheep. The technology of hand knitting requires not just the yarn and needles themselves but also a technological practice, the knowledge of how to use the yarn and the needles to produce a knitted garment. My mother taught me the basics of how to knit when I was a child, and then I learned more from patterns, instructional books, and the occasional online video tutorial. In *The Real World of Technology*, Ursula Franklin describes such a practice as a "holistic technology," one in which a single person understands and has the skills and autonomy to carry out every step of the process, closely related to the idea of artisanal labor. Even today, there are people who raise sheep, turn their fleece into wool, and knit the wool into garments.

But at the edge of my Faroese shawl is a row of buttons made of matte black plastic. There is no single human on the planet who can fabricate one of those buttons from scratch, or who even fully understands all of the pro- cesses and techniques involved: extracting the oil, refining it, overseeing the chemical processes required to transform it into a polymer, deciding on ad- ditives like pigment and opacifiers to make up the final formulation of the plastic, designing the shape of the button, setting up the injection molding process . . . just a small subset of the skills and knowledge needed to create a plain plastic button. Over the course of the past century or so, most of the technologies behind the material realities of modern life in industrially developed regions became far too complex for an individual to encompass, whether as tiny and commonplace as a plastic button or as huge and compli- cated as a data center or a nuclear power plant. These artifacts and techno- logical systems are a product of social and organizational structures that enable people with specialized skills to work together. Standards, in particu- lar, modularize and black-box all the different technologies and subtech- nologies, making them manageable and interchangeable even without a complete understanding of their internal details, which in turn enables the advanced cooperation that makes larger technological systems possible. This is a societal shift that's hiding in plain sight: most technological sys- tems, including mass-produced material goods, are now products of collec- tive, rather than individual, intelligence. As a general rule, if it didn't already exist about a hundred years ago, it's not something that a single person could fully understand or create today. The fabric of daily life is created from tech- nologies that have outpaced our ability as individuals to fully understand or recreate them. And that includes the technologies that are the interfaces to infrastructural networks.

Size, Scale, Extent, and Complexity

My car is considerably more technologically complex than my bicycle, incorporating a much wider set of subtechnologies in its function. Pretty much every element of my bike (the frame, the wheels, the brakes, the gearing and transmission, even the bell) has an analogous system in my car, but the converse isn't true: my bike doesn't have an internal combustion engine, or a catalytic converter, or an air conditioner, or a stereo. Cars can go farther and faster than bikes because they generate more motive power than bikes do, made possible by consuming gasoline as a source of energy, rather than relying on the limited capacity of my leg muscles. More energy to move them means that cars can be bigger and heavier than bikes, which in turn means they require far more energy to manufacture. This gives us a hint as to one of the drivers of increasing technological complexity: access to energy.

A graph of the total exogenous energy usage of humanity (that is, energy from all sources outside our own bodies) over time is flat until about 1800, after which it becomes roughly exponential, starting slowly and then rising more and more steeply. The largest share of that growth comes from the increasing extraction and combustion of fossil fuels. The very earliest commercial steam engines were used to pump water out of coal mines, which made it possible to dig out more coal. My colleague Ben Linder pointed out that it was a physical mechanism for exponential growth: these mechanical pumps could reliably turn a small investment of energy (from the coal burned to produce the steam that powered them) into a much larger amount of available energy (produced from the increased amount of coal that was mined). As more energy became available, it was used to create and power more, bigger mining equipment, first to extract yet more coal, then oil and natural gas. The energy produced was in turn invested into building dams for hydroelectric power, into nuclear reactors, into solar panels and wind turbines, resulting in the continuing exponential growth of total energy production and use. A system where much more energy comes out at each step than was put in is behaving more like the runaway chain reaction of a nuclear bomb than the controlled simmer of a reactor core.

This exponential growth in energy usage both motivated and literally powered the increase in technological complexity, not just in energy infrastructure but everywhere. It used to take whole crews of longshoremen to unload and reload ships; now that work is done by a small handful of crane operators, mediated by a technological system that makes it possible to manage the energy required to move and stack ten-ton shipping containers. This technological complexity, energy usage, and network growth is self-reinforcing, resulting in systems that function in ways that are not only, in the words of the song by Daft Punk, harder, better, faster, stronger, but also bigger, farther, and serving more people. And more functional, more reliable, and more specialized—but also more centralized and more extractive.

Moving Bodies and Bits

Shipping containers created large-scale global networks for the transportation of goods, but the second half of the twentieth century also saw the rise of two other global networks: civil aviation for people and telecommunications for information.

My parents emigrated from India to Canada in the late 1960s, and their arrival coincided with a time of rapid (and continuing) expansion of air travel. Immigration was no longer a one-way trip. They could take their young family home to visit extended family in Bhopal and New Delhi on a fairly regular basis, and they did. I don't even remember the first time I was on a plane, because I was too young. I do remember that, before I learned that the earth was round, I had already deduced from my travels that all the different places in the world were stacked on top of each other, separated by clouds like layers in a cake. This meant that you had to go up or down to get to them, which explained (to five-year-old me, at least) why you had to fly.

Childish explanations aside, the growth in civil aviation—the jet age—was kicked off by and takes its name from the development of robust, reliable jet engines, which could be made larger and more powerful than their piston-engine predecessors. Half a century or so later, a modern passenger jet plane, like an Airbus A380, exemplifies the trajectory of technologies toward larger scale, more complexity, and more energy usage. It's a staggeringly complex artifact in its own right; the design, manufacture, operation, and maintenance of a plane like this is only possible because of organizational structures in which hundreds or thousands of people with vastly different skills and knowledge collaborate and assemble technologies into something that reliably stays up in the air. As with container ships, modern passenger jets are bigger and can go farther than their predecessors: an Airbus A380 holds 525 passengers and has a range of just over nine thousand miles, which means it can fly nonstop between Hong Kong and New York City. This is made possible by access to energy in the form of fossil fuels. In the case of a fully fueled A380, that means more than 250 metric tons of aviation kerosene, or somewhere north of 80,000 gallons. On a per-person, per-mile basis, though, an Airbus A380 gets better gas mileage than I do on my daily commute in a compact hatchback. A bigger plane with a longer range that carries more passengers can do so more efficiently, and when fuel is a large chunk of operating costs, that can mean more cheaply and more profitably.

The cost of passenger travel fell and the global middle class continued to grow, resulting in a roughly fifteenfold increase in the number of air passengers worldwide since 1970, to about 4.5 billion passengers and something like 40 billion commercial flights in 2019. Planes don't exist in isolation, and this growth in air travel meant the commensurate growth and development of everything else that makes it possible for them to move safely around the world. Commercial aviation leverages all the systems that I've described—global navigation, coordinated and precise timing, standards, and international cooperation around every element of travel. This increase in the accessibility of personal mobility in the last part of the twentieth century and into the twenty-first—not just the proportion of people who travel, but how often and where to—is part of what we mean when we talk about living in a globally connected world, with perspectives informed by visiting other places, by being or living alongside immigrants, and by international trade.

Completing the trifecta of planetary-scale infrastructural networks is telecommunications. Throughout history, the transmission of information over long distances largely required the transportation of atoms: messages were necessarily physical, like letters. The international mail system for global communications—reciprocal agreements to deliver mail regardless of which country's post office was paid for the stamp—dates back to 1874. The advent of telegraphy made it possible for communications to be pure signal, mediated by electrons transmitted along wires and undersea cables. This was followed by telephony, and then radio communications, first as telegraphy and now, a century later, as wireless networks. The final few decades of the twentieth century saw the rise of high-capacity long-distance telecommunications cables for voice and for data. With these connections in place, the decentralized, packet-based routing protocols of the Internet facilitated global communications. In many places that were previously poorly served by telephone service, wireless telecommunications have entirely leapfrogged landlines, with near-complete penetration of cellular telephony and rapidly rising access to mobile data. The rise of digital telecommunications has made it possible to largely decouple sharing information from sending physical materials in the form of letters or other documents. This was recently driven home to me when a colleague was talking to our undergraduate engineering design class about a project for bicycle couriers, and one of our first-year students put up their hand to ask, "What's a bike courier?" Despite—or maybe because of—the way that the vast majority of my communications are now digital and ephemeral, I'll probably always have a place in my heart for material, enduring postal mail. It still remains remarkable to me that we have a publicly administered, globally distributed, fully reciprocal supply chain for the one-to-one delivery of physical objects.

Learning to See the Sublime

Helen Macdonald, the naturalist and historian who talked about the "green blur," also wrote about the times when their "attention has unaccountably snagged on something small and transitory" in the natural environment and how these elements serve as tiny portals, as long as the observer possesses a "sufficient quality of attention to see them at all. When they occur, and they do not occur often, these moments open up a giddying glimpse into the inhuman systems of the world that operate on scales too small and too large and too complex for us to apprehend." The moment of noticing sparks Helen's remembrance and awareness that they are surrounded by and embedded in the natural world, of the sea and sky and the land and the ecological systems that connect them through time, revealing the "numinous ordinary."

Their account comes strikingly close to how I feel when I look at the three-letter airport code on a boarding pass and consider how a bar code grants me passage to the entire world. Or when I press a switch and think about how one wire goes to the lightbulb in my bedside lamp and the other wire goes hundreds and hundreds of miles, to a dam in northern Quebec or to a nuclear plant or to a solar farm. Or when I look at the blue dot on my phone and realize that it's listening for signals from satellites and unimaginably precise clocks while I'm standing on the ground. The writer Tim Carmody has called this the "systemic sublime"—a parallel to Helen Macdonald's numinous ordinary for human-created systems instead of the natural world.

All of these systems interact and are entangled with one another, not just in technical ways, like how a washing machine uses both electricity and water, but in ways that are mediated by enormously complex social, legal, and regulatory systems. Many books have been written about water in the western United States, and many professional lives spent understanding the legal frameworks that have accreted over the course of half a millennium since European colonizers began to settle the region. The food-water-energy nexus means that it's not possible to consider any of these systems in isolation. They are entirely situated in a sociotechnical context that includes the technical and environmental considerations and these human decision-making systems.

In many ways, this is what we think of as being "modern," particularly as residents of a wealthy, economically advanced country. It means that our way of life isn't constrained by the resources that are found in the immediate environment, and that everything we do is embedded in and enabled by global systems of energy, trade, communications, migration, food production, and supply chains. Because the existence and nature of these systems shape the kinds of things we can do and how we do them, the possibilities that are available to us on a daily basis, how we feed ourselves and stay warm or cool or move around or have water for drinking and hygiene, that means that *we literally don't know how to live without them.* If you're reading this book, you've almost certainly had the feeling of glimpsing the systemic sublime, the *human* systems of the world that, in Helen's words, operate on scales too small and too large and too complex for us to apprehend, vast and seemingly unknowable. The brain-exploding insight, the one that we all live with whether we acknowledge it or not, is this: for a large fraction of humanity, the ways we meet our most basic needs, the ways we carry out the most routine and quotidian activities of our daily lives, and the material culture in which we are embedded are all *beyond our capacity to understand, or even fully apprehend.*

But this isn't where we stop.

It's easy to feel like we have no control or agency over these systems, that if we can't comprehend how they work then we can't predict their behavior, much less change them. But unlike the "numinous ordinary," which describes the natural world that has been evolving for many millions of years, the systemic sublime is about human-created systems. There is nothing natural or inevitable about the way they work.

Shaping the Systems

Other than roads and water systems, most of the networked infrastructural systems that we interact with every day have only existed in recognizable form for about two centuries, since the Industrial Revolution, because their operation depends on abundant and reliable sources of energy, harnessed by increasingly complex technologies. The truly global nature of supply chains, mobility, and telecommunications has been in place for less than a human lifetime.

First of all, this is remarkable: as a species, we've learned to cooperate well enough that we marshal the resources to provide for the daily needs of a substantive chunk of the world's population, in the billions, and we do it well enough that many of those people don't even have to think about how it happens. But just like it's easy to never think about a specific infrastructural system until it fails, it's easy to never think about infrastructural systems *as a whole* unless we are in a position to recognize the ways they restrict what we can do or to observe the ways in which they cause harm. What's *also* changed within a human lifetime is that even the most privileged users of these systems—I'm absolutely one of them, of course—are now unavoidably bumping up against these negative effects, especially the understanding that the combustion products of fossil fuels are driving anthropogenic climate change. It's this realization—that we want to interact with these systems differently—that immediately leads to that creeping or even overwhelming feeling that we do not understand them and that they are too big and complicated and entangled for us to have any control or agency over how they work.

Because these infrastructural systems are about moving resources through a network, changing with time and incorporating feedback loops, they can behave in ways that aren't immediately clear, or that are even counterintuitive. But that also means that human choices shape these systems every day, day after day—they change and respond over time as people interact with them. Roads are a familiar example. We might think of them as inert and passive, channeling the flow of traffic like aqueducts carrying water. A bigger aqueduct can move more water, but if the same logic is applied to highways—adding extra lanes in a bid to reduce traffic congestion—it doesn't work. Instead, the number of vehicles quickly increases to match the carrying capacity of the new road—a phenomenon known as induced demand—until the congestion is just as bad as before. Flowing water molecules are channeled blindly, governed by the mathematical equations of fluid dynamics, but traffic is the result of thousands or tens of thousands of individual drivers making decisions every day about how smooth or aggravating or long their commute will be, and decisions over the course of years about where they live and work.

Humans aren't really great at understanding and interacting with these kinds of systems, largely because they can behave in ways that are far more complicated than simple cause and effect, even if it's just water molecules. A five-year-old understands that if you put a stopper in the drain and turn on the faucet, a bathtub will first fill and eventually overflow—it's a straightforward, one-way network (it's also why sinks and tubs commonly have overflow drains). But connect together a bunch of bathtubs with a bunch of pipes—each with its own faucet and drain, all connected so that the water that leaves one bathtub might go to a different one—and the system is not only more complicated but also its behavior becomes much harder to predict, especially in response to changing circumstances like turning off specific faucets or stopping up particular drains. Where will the water go? Which bathtubs will drain entirely, and which will overflow?

Just because the ways in which they behave aren't obvious doesn't mean that they aren't knowable. In her essay "Holy Water," Joan Didion described her visit to the Operations Control Center for the California State Water

Project in Sacramento. It's one of several agencies that move water around the region, between reservoirs, through dams, and along aqueducts, and the operators have a nuanced understanding of the system, a huge, complicated, constantly changing, high-stakes set of networked bathtubs. Grid operators do the same kind of work with electricity, matching production and distribution to demand on a moment-by-moment basis. Induced demand has been carefully studied by transportation researchers, who've demonstrated that however counterintuitive it seems, the correlation between more and wider roads and increased traffic is predictable and quantifiable.

The researcher Donella Meadows pioneered the field of studying these types of networked, time-dependent systems, which describes much of our infrastructure. She points out that deterministic, analytical tools can't necessarily capture their behavior. Rather, like complex ecological systems, they reward long, careful observation. Sometimes that understanding comes just from paying attention as we go about our daily lives. I've driven the easternmost stretch of the Massachusetts Turnpike—Interstate 90, between the city of Boston and the ring road of Interstate 95—most days of the year for the better part of two decades now. I know the behavior of traffic on that highway like a riverboat captain would have known the Mississippi—I know where the current runs fast, where it runs slow, and the dangerous spots where it changes from one to the other. I know how that flow changes in the summer, in the winter, in rainstorms and Red Sox games, and during a global pandemic. Observational understanding over time is a powerful tool to get a handle on how these systems work.

Didion wrote that visiting the operations center made her want to be, herself, the one who opened the valves and closed the gates to move water all around the state. Humans are really, really oriented toward storytelling. Given that we live inside our own heads, walking around, doing things, and having things happen to us—a narrative, in other words—it's not surprising that stories that involve individuals progressing through time and moving through space would be part of the basic human tool kit for making sense of the world. We are solipsistic narrative-generating machines. But when you're interacting with infrastructural systems, individual considerations often don't count for much. My friends joke about how I have "good airline karma" because of how consistently lucky I've been at avoiding flight delays and cancellations. A good chunk of it is improbable and random, for sure, but it's also rooted in that deeply internalized understanding that civil aviation is a set of giant interlocking systems, and I'm only the tiniest part of it. Having some understanding of how airlines work has served me in good stead, but the first step is learning to see collective systems as, well, *collective* systems. I'm part of it as a passenger, but so are the staff, whose behavior and flexibility within the system is likely even more constrained since their livelihood depends on it. I've only witnessed a "Do you know who I am?"–style meltdown at a gate a couple of times. They're rare because most adult humans, even when they're upset, do understand that their needs or desires don't outweigh those of a passenger jet full of people. It's the same idea as "You're not stuck in traffic, you *are* the traffic." We can consciously sidestep the narrative solipsism of thinking we must be the hero of the story (or that *anyone* is) and the attendant fallacy that the systems are there to serve us individually or specifically, rather than as part of a collective. It's even more evident when they fail in some way—weather, mechanical issues, scheduling glitch—because it affects that entire passenger jet full of people and cascades onward from there.

Alongside the idea that individual decisions result in collective outcomes (as with traffic) or that individuals have relatively little power in collective systems (like when they're late for a flight), is the idea that individuals can indeed act collectively, and that we can use new tools to do so. One of the most powerful examples I've seen of this came on September 6, 2022, several days into a heat wave in California. As predicted, demand for electricity had set an all-time record for the state, early in the late-afternoon peak, and was still rising. Just before 5:30 p.m., the power grid went to the highest emergency level: nearly every scrap of reserve capacity available in the system was spoken for. If demand exceeded supply, it would mean rolling blackouts. I was keeping an eye on the grid operator's page and something odd happened. Just before 6 p.m. local time, the demand suddenly *fell* sharply. What I hadn't realized, from my desk on the other side of the country, was that an emergency text message had gone out to Californians, asking them to take action: to turn off anything nonessential and to turn down their air-conditioning. And they did. Within minutes, demand had dropped by the equivalent of nearly a million homes' worth of electricity and the crisis was averted.

The Human Systems behind the Networks

Energy, networks, standards, scale, time: these are the building blocks of infrastructural systems, as systems. Not just each individual type of system, or the elements of it that we can see and interact with, the power outlets and highway overpasses and bridges and cellphone towers. Rather, we now have a conception of how these systems are complex technologies in their own right, rooted in specific features of the material and technological world, that they are part of global networks resulting from global cooperation, that they use energy to move resources to where our bodies are, and that they are dynamic and interconnected.

The systemic sublime is about more than just technology—or rather, technology is more than just the physical systems. These are human systems, created by and for humans to serve particular needs and desires, and to meet them in particular ways. They are intrinsically collective systems, which call for the ultrastructure of collective decision-making; this includes governance and regulation, agreements and policies and treaties. Why do our systems of infrastructural provision function the way they do? Why were they built out the way they were? Economic imperatives, political structures, and social norms all had a part in shaping them.

Five

THE SOCIAL CONTEXT
OF INFRASTRUCTURE

I have a deep affection for a lot—*a lot*—of charismatic infrastructural mega-structures. When I transit through Chicago O'Hare or London's Heathrow Airport, it feels like I'm in the beating heart of the world, at the nexus of worldwide systems and thousands and thousands of human stories. I still feel warmly toward the Pickering Nuclear Generating Station where I spent so much time at as a child. The combination of the vast scale and ornate detailing of the Hoover Dam, set amidst its arid, austere desert surroundings, is brain-melting. I adore the appealingly calm quiet of the paired hydroelectric power plants set inside green and forested Niagara Gorge, several miles downstream of the thunderous falling water and throngs of tourists at Niagara Falls. But if I did have to name a favorite, one infrastructural installation does stand out. It's simultaneously enormous and almost entirely hidden, an elegant engineering project that serves what at first seems to be an almost whimsical human purpose.

Snowdonia National Park, in north Wales, is idyllic, all green hills surrounding blue glacial lakes. Long-standing farmhouses, ruined castles, and modern buildings alike show off the gorgeous locally quarried slate, the mining of which grew into an important industry in the nineteenth century (the region recently became a UNESCO World Heritage Site because of the historic mining facilities). A Victorian-era railway carries tourists to the top of Mount Snowdon to take in the view. If you're lucky enough to be there in May, like I was, the charm of the landscape is further enhanced by absolutely adorable gamboling baby lambs *everywhere*. But I was there to see something that had been made deliberately, carefully, and beautifully invisible, concealed inside a giant cavern carved out of a mountain. Like a supervillain in their lair, it gathers power to itself, usually under cover of darkness. It's just that the power, in this case, mostly lets people watch TV and make tea.

Snowdonia is the home of Dinorwig Power Station, popularly known as Electric Mountain. Together with some friends who live in Northern England, I had made my way to the visitor center next to the train station to join a guided tour. It began with an introductory video describing the facility, filled with claims about its size: "the largest cavern in Europe!" "two Empire State Buildings high!" and "three Blackpool Towers high!" (Without the implied conversion factor, that last comparison would have been opaque, but my accent from the other side of the North Atlantic made it clear that I was an outlier.) We were all made to stow our phones and cameras in lockers before boarding a minibus that trundled through the Welsh countryside for a few minutes, passed through several security gates, and then went down a tunnel deep into the mountain itself, Elidir Fawr.

Conventional electrical grids largely operate with just-in-time energy production, where the exact demand is estimated on a minute-by-minute basis, and electricity is generated and allocated accordingly. The whole system is continuously and carefully monitored because failing to match the demand with adequate supply leads to brownouts or even blackouts. Most of the demand is predictable. There's a daily cycle of electricity usage, commonly known as the "duck curve," tracing the profile of a duck's back and neck over the course of the day. The duck's tail is the low morning peak, as people wake up and start their day, the steady demand through to the late afternoon traces its back, and then a sharp rise in demand corresponds to the back of the duck's neck, peaking at the top of its head in the evening, as everyone turns on lights, makes dinner, does chores, or has the TV on. Finally, there's an overnight low in aggregate demand while most people are sleeping. The exact pattern and amount of electricity usage on a given day can be affected by other factors, like holidays and the weather. For unusually high loads, electricity is often supplied from "peaker plants," designed to be brought online as needed and which run on fossil fuels, usually natural gas. While expensive to operate, the electricity they produce commands a premium price because of the high demand. A typical use case for peaker plants here in the northeast U.S. is during hot summer afternoons on workdays, when air conditioners are running full tilt. Electric Mountain is a kind of peaker plant, but part of what makes it special is that it can go from idle to producing power in minutes, or in as little as *fifteen seconds* from standby mode.

Construction at Dinorwig started in 1974 and the facility came online ten years later. In the control room, operators ensure an adequate supply of electricity in the grid by carefully monitoring their inputs, ready to act when needed. One of those inputs is watching television. Electric Mountain was conceived and designed in the late 1960s, and that period through the 1990s was the heyday of synchronous, broadcast television, where nearly everyone in the UK might be watching the same TV series finale or sporting event at the same time. And then, when the show was over, a substantial fraction of the viewing populace did the same thing, in a phenomenon known as "TV pickup": they reliably and immediately plugged in a power-hungry electric kettle to make tea. Alert operators could open a set of valves the instant the television broadcast ended, sending a spike of power into the grid in less than the time it took for viewers to get up from the sofa, stretch, and walk into the kitchen.

This rapid response is possible because Electric Mountain is a pumped-storage hydroelectric station. From the outside, the installation consists of Elidir Fawr and two placid, slate-bottomed lakes: Marchlyn Mawr, high on the flank of the mountain, and Llyn Peris, nestled in the valley between Elidir Fawr and Mount Snowdon near the visitor center. Inside the mountain, they're connected by a vertical shaft. The half a kilometer or so of altitude between the two lakes allows them to serve as upper and lower reservoirs. When an operator opens the valves at the top, water from Marchlyn Mawr flows in and falls straight down onto the turbines of six generators lined up in "the concert hall," the cavern in the mountain that's big enough to comfortably house St. Paul's Cathedral, and then out to Llyn Peris. At full flow, enough water to fill a twenty-five-meter swimming pool passes through the turbines every second (appropriately enough, the tour guide described this as "1.5 million cups of tea") but even at this rate, the artificial waterfall inside the mountain can generate electricity continuously for six hours before Marchlyn Mawr is depleted.

All of this is impressive enough—by being able to produce this power at a moment's notice, Dinorwig dramatically increases the reliability of the UK electrical grid. But the real elegance, to my mind, is what happens next. When demand is low, there's surplus capacity available because of baseload power generators, like nuclear plants, that supply a near-constant amount of electricity. After all those television-watching Brits go to bed, the generators are run *backward*. Instead of the falling water spinning the turbines to generate electricity, power is fed *into* the generators, causing the blades to rotate. Now it's an electrical motor, and the turbine works as a pump, lifting water from the valley all the way back up the shaft to the higher reservoir and so resetting the system. It isn't perfectly efficient—that would violate one of the laws of thermodynamics, which mandates that something has to be lost in the process—but it's pretty good, with about 75 percent of the energy available for reuse. Electric Mountain makes engineering sense as a way to deal with unpredictable spikes in demand, but it also makes economic sense because it can time-shift energy usage from expensive high-demand times to cheaper low-demand periods, arbitraging the price difference. What's more, power from pumped-storage hydro has a much lower carbon footprint than conventional peaker plants.

Dinorwig sets my engineering heart aflutter not just for its scale, although size does indeed matter, but because of its undeniable elegance, the way it solves a problem by using the available resources—mountain lakes! low overnight demand!—cleverly and parsimoniously. And I admit that I do find the idea of a power station that was built, even in part, so that an *entire country* could turn off the telly and all go make themselves a nice cuppa to be kind of hilarious. Since the advent of first video recording and then streaming, television viewing has become much less coordinated, and so TV pickup is now much less pronounced. But Dinorwig is becoming more useful and valuable, not less. As renewable electricity generation increases, the inherent variability of sources like solar and wind means that more grid-scale energy storage is necessary in order to make sure that there's always electricity available to meet demand. A pumped-storage hydroelectric power station like Dinorwig is effectively a giant rechargeable battery.

But impressive though Electric Mountain is as an engineering accomplishment, that's not what I think is the most astonishing thing about it. It cost nearly half a billion pounds to excavate the inside of Elidir Fawr and construct the facility, making it the largest civil engineering contract ever awarded by the UK government at the time, and it required all the cooperation around technical expertise and standards that we discussed in the last chapter. But in a larger sense, what really makes Dinorwig and similar megafacilities remarkable is the level of social, political, and economic organization that underpins the decision to build them at all.

Ten thousand years ago, give or take, *Homo sapiens* began to produce food via agriculture, which led to permanent settlements. About five thousand years ago, the urban revolution began. Many people living in proximity to each other made more complex forms of social organization possible, and the rise and spread of cities led to the generation of surplus beyond mere subsistence, which in turn enabled new modes of human activity. Then, from about five hundred years ago, humans suddenly began to get a lot richer. How suddenly? Economic historians have quantified it by estimating the per capita gross domestic product in constant dollars. It's basically a flat line for fifty thousand years and then, starting in Europe around 1600, it starts to rise, getting steeper and steeper. Economist Deirdre McCloskey gave it a name: the Great Enrichment.

The *what* of the Great Enrichment can be stated simply: humans started to get much, much better at cooperating in ways that led to the creation of economic surplus. People learned and were incentivized to work together to produce more and more wealth, far beyond what we had ever been able to do as individuals, or neighbors, or kinship groups, and this became part of what led to systems of standards, networks, and technological cooperation. As we've seen, these cooperative structures have grown and grown until they span the globe. The Great Enrichment is why my daily life isn't focused on caregiving and subsistence farming.

The *why* (or rather, the *Why then?* and *Why there?*) of the Great Enrichment remains a contentious area of debate among social scientists, economists, and historians. One argument is that there was something special about Western Europe: some way of thinking that fostered collaboration between peers that was coupled with the idea that no matter where one stood in society, having more wealth was desirable and possible, with the Scientific Revolution and the Enlightenment as likely contributors. The growth in wealth is almost certainly linked to private ownership of resources, particularly the idea of enclosure, which was the creation and extension of property rights over open fields and common land which allowed for the exclusion of non-owners. The invention of the corporation, a way for unrelated individuals to collectively invest, might also have been a factor. No matter how the Great Enrichment got started or where, it enabled the exponential growth of accumulation of capital, the reinvestment of wealth to rapidly produce more wealth. Just like cultures in a petri dish are often quickly dominated by whichever microorganism can replicate and divide the fastest, it seems likely that the first society that got traction on this process, regardless of the initial impetus, would have a significant economic advantage. But there's no question that the uneven geographic distribution of this rapid growth in wealth—a distribution that's still apparent today—was extended and accelerated by European colonialism.

Even if we can't isolate the initial cause of this sharp increase in economic growth, we are all living in a world shaped by its effects. It's especially apparent at facilities like the Dinorwig Power Station, because they're a manifestation of a collective economic surplus being put to work. There's a familiar pattern for infrastructural systems: they often require a significant investment in the form of capital costs, paid up front, before the system can be used for the very first time. Think about New York City building out its dams, reservoirs, and aqueducts, all of which had to be in place before the first drop of water came out of a Manhattan tap.

The economic argument for building systems like Dinorwig or a municipal water supply is that the large initial investment pays off: once the system is built, it makes provision cheaper for a long time afterward. Public infrastructure is often funded with bonds because the large upfront cost can be balanced against their steady payout rate over time. This seems mundane, but it's actually remarkable. First, the construction of systems like these is predicated on the belief that there will be a stable community in place to benefit from them, typically for decades to come. They're an investment into a vision of a *shared* future. They are, by their nature, about continuity and time. Next, the economic argument for their value is strengthened by universal provision because the capital and operating costs can be amortized over more users. Finally, membership in that social collective is primarily based on *where* you are, not on kinship or social class or anything else. One definition of politics is just that it's what you get when a group of people have an ongoing relationship with one another that they can't easily walk away from, so they need to figure out ways to work together. One of the most profound and sustained ways in which we are in a relationship with other people is just by living near them and sharing the physical systems we use to meet our needs. A focus on the engineering and technical considerations of infrastructural systems can tell you that the most efficient thing to do is to build a centralized water treatment plant and a network to distribute it. But this can only be achieved—planned, designed, and paid for—collectively. A pumped-storage hydroelectric plant or a municipal water system is a physical instantiation of an ongoing relationship between the people who share it.

Infrastructural systems become part of the way that communities generate a growing wealth surplus through economic growth, engendered by cooperation. Building out a municipal water supply or a public power utility is a way to leverage economic surplus by reinvesting it into even more cooperation and surplus, over the course of decades to come. This sustained, mutually beneficial relationship between people who are near each other, mediated in part through infrastructural systems, can itself become a factor in the uneven geographical distribution of wealth.

Infrastructure as Agency, Revisited

When a pandemic shut down his college, a twenty-three-year-old student returned to his family home and found productive ways to occupy his time for the better part of two years. The college was Trinity, at Cambridge University, the student was Isaac Newton, and the independent research in optics, gravitation, and calculus that he carried out at Woolsthorpe Manor during the enforced leave from his coursework would result in 1666 being known as his "year of wonders."

The residential engineering college where I teach was similarly shut down in 2020, as we sent our students off to their own homes to help keep them safe during the height of the global coronavirus pandemic (although their coursework continued, in modified form, thanks to modern telecommunications). Working at home in another Cambridge, four centuries and three thousand miles removed from Newton's, I remember hearing about his *annus mirabilis* during the Black Death, occasionally presented as an "inspiring" example of being productive during a global health crisis. It's rarely mentioned in the popular accounts of his year, but young Isaac Newton had a cohort of servants in his home to keep him warm and fed and who made sure there were tapers or lamps for him to see his experiments and equations by. It occurred to me that most of these day-to-day needs were also met for me, not by a household staff but by infrastructural utilities.

I did not, of course, revolutionize our understanding of physics from my home during the pandemic because I'm not an Isaac Newton, and I probably don't know one personally. But on a planet of eight billion people, there are Newtons out there, and Beethovens, and Cervanteses, and da Vincis. Most of them haven't been able to live up to that potential. And while I name-check Newton and Beethoven and Cervantes and da Vinci, they all surely had female peers who never had the opportunity to go to Cambridge, much less spend their time drawing, carrying out experiments, or composing music. By and large, those women would have been kept busy making sure the needs of the people around them were met, whether it was what they wanted to do or not. That remains just as true today for the large fraction of the world that doesn't have secure food, water, sanitation, or housing. It's often framed as humanity "losing" a potentially brilliant scientist or composer, but it's about *everyone* having more agency in their own lives, and the possibility to develop their individual potential. In his book *Development as Freedom*, the Nobel Prize–winning economist Amartya Sen wrote that we "generally have excellent reasons for wanting more income or wealth. This is not because income and wealth are desirable for their own sake, but because, typically, they are admirable general-purpose means for having more freedom to lead the kind of lives we have reason to value. The usefulness of wealth lies in the things that it allows us to do—the substantive freedoms it helps us to achieve."

If, as Jean-Paul Sartre wrote, "Hunger is precisely in itself the demand to be something other than a belly," then the most basic freedoms result from having your bodily needs met: to not remain hungry, or thirsty, or too hot or too cold or too fatigued or too sleepy, and not to be sick or in pain. Sen's research on the world's poorest people considered how they might get those basic needs met: freedom *from* having those needs. But as more resources become available, that opens up freedoms as a space of possibility, the freedom *to* lead the lives we value. Artificial light provides a good example. Being able to function after the sun goes down or in an enclosed space is an immediate extension to one's personal agency. If money is a general-purpose tool to achieve freedom, then using it to buy more lumen-hours of light is among the most direct paths there.

This is the role that infrastructural systems play in my own life, and how they provide me with agency—they are a means of freeing up my time, energy, and attention. I could probably marshal the skills and resources to build out a home that mostly stands apart from at least some of these systems—off the grid, with its own water supply, a septic system, solar panels, maybe a high-efficiency stove. But then I'd have to operate and maintain these systems, and that means that I would have notably *less* freedom in my life, not more—in a very real way, it'd just be a higher-tech version of what my female ancestors had to do every day. I also know enough about the systems that serve my home, and the specialized skills of the people who keep them running, to know that nothing I could do myself would ever result in anywhere near the level of quality and reliability that allows me to mostly not think about them. On a day-to-day basis, my personal agency doesn't come from money *per se*—it mostly comes from living in a place that's served by high-quality infrastructural systems. But the bigger point is that access to energy and resources through shared infrastructural systems frees up not just my own life but those of everyone around me. The real difference between me and Isaac Newton isn't our own lives, which in some ways are actually remarkably similar. It's the difference between our lives and those of the other people in his home, a class of subservient humans, economically or even forcibly coerced to household labor. Were I in Woolsthorpe in 1666, my gender would almost certainly have put me in with the servants, not with the scholars.

If you're privileged enough that you normally have access to this full-stack infrastructure, you're likely accustomed to these systems only ever increasing your agency. From the perspective of the richest people on the planet—of whom I certainly am one, as part of the global 10 percent—the relationship between infrastructural systems and freedom always becomes most clear when something fails: the power goes out, a water main breaks, a global pandemic means we don't travel or even commute. Our baseline shifts and our lives are suddenly much more constrained because our attention and efforts are necessarily focused on how to see when it's dark, how to stay warm, what we'll drink, how we'll dispose of our waste, how we'll cook our food, or how we'll stay connected to others.

Market Goods and Network Monopolies

Near the beginning of *Mad Max: Fury Road* (one of my all-time favorite movies), the warlord Immortan Joe, whose citadel is built on top of an underground spring in the postapocalyptic Australian desert, opens a valve and water gushes out onto a desperately thirsty crowd below. "Do not, my friends," he roars over a loudspeaker, "become addicted to water. It will take hold of you and you will resent its absence."

Water, sewage, electricity, mobility, communications—infrastructural networks provide resources that don't behave like consumer products or services. They aren't market goods that you can elect to purchase or not, now or later, from your choice of sellers. Buying a car is different from having access to a highway. Having electricity service is not like getting a haircut.

For a start, infrastructural systems are often used to provide survival needs. There's a reason why control over access to water, whether physical or economic, is a staple of dystopian science fiction. Water isn't a consumer good, and potable water for drinking, as well as for sanitation and hygiene, is simply not subject to the usual laws of supply and demand. You can't choose not to drink water if it's too expensive. You can't even put it off in the hope that the price will drop a week or a month from now. In the depths of a New England winter, the means to heat one's home is a necessity, not a choice. Even though it's not directly required for survival in the way that water and heat are, we use electricity—raw, versatile energy delivered directly to where we need it—to provide for a wide range of needs. It's part of why blackouts are so dangerous: lives are lost as a direct result of power outages in homes, such as when people have medical conditions that require the use of at-home respiratory assistance. A power failure during a winter ice storm or one that knocks out air conditioners during a heat wave can exacerbate medical is- sues. These indirect effects can be revealed by excess-death analyses, in which the deaths during and immediately after a power failure are com- pared to a baseline rate. Some of the lethal risks of blackouts result from at- tempts to cope with the lack of grid power—in recent years, more people in the U.S. have died of carbon monoxide poisoning from emergency genera- tors in the aftermath of major hurricanes than from the storm surges and flooding themselves.

Even if the services provided by infrastructure networks aren't survival needs, they can still be nondiscretionary and non-fungible—it's difficult or impossible to substitute something else. Communications is a good exam- ple: it's not going to do me any good to send you a fax over a phone line if you're expecting me to fill out an online form. During the COVID-19 pan- demic, the shift to online education meant that reliable access to the Inter- net suddenly became a nonnegotiable requirement for classwork. If you need a car to get around your neighborhood, you can't choose to fill up the gas tank of your car next week if it's empty today. If you commute by public transit, you can't "buy less" of a subway fare if the price goes up. Because our use of shared infrastructural networks is rarely at our individual discretion, the usual economic rules of supply and demand mostly go out the window.

We also usually don't have much choice in *which* infrastructural systems we use. This might be because of limited access, as exploited by Immortan Joe, whose citadel was the only source of water for the people living nearby. Usually, however, it's because of the way that networks function. Their efficiency comes with some drawbacks.

It's a cliché in the U.S. that everyone hates their local cable company, which is now often their Internet provider. In many cases, there's only one provider in a given region, because that sort of "natural monopoly" is a consequence of the way that networks work. They're often cost-effective because, while they might have a large upfront cost to build out, the cost to add nodes is relatively low. But building out a new network doesn't make economic sense unless the provider can recruit enough customers to pay for their service. The problem is that a second network in the same area, much less a third or a fourth, might cost nearly as much to build out as the first but it could, at most, only get a fraction of the total paying customers. The network provider has a significant first-mover advantage—as they build out their user base, they are also creating a barrier to entry for new competitors. The usual end result of this is just a single provider of the service in a given area: a monopoly that arises "naturally" from the characteristics of networks, especially physical ones. That means that one of the most commonly extolled virtues of a free market—that companies compete with each other to offer the most appealing service at the most reasonable price—isn't in effect.

Private companies that have a monopoly on a necessary service behave in pretty much the way that you'd expect they would. In the best-case scenario, customer service is terrible, which leads to endless griping and jokes about avoiding interacting with cable companies at all costs. But what's worse, investor-owned providers can decide to increase their profit by raising prices or downgrading service, secure in the knowledge that their customers can neither choose a different provider nor do without—the classic reason why monopolies are considered to be at odds with the best interests of consumers. One response to this is oversight: utilities, whether private or public, tend to be heavily and carefully regulated at all levels. Where I live, for example, private companies that provide natural gas for home heating are forbidden from shutting off service to any household, for any reason, in the wintertime. But this also means that consumers are at the mercy of those regulations and whether they have teeth. Oversight and enforcement vary widely with the local regulatory environment because there's always a tension between private providers and regulators.

A major strategy for avoiding the moral hazards of network monopolies in private hands has been for them to be publicly owned or nonprofits from the start, like the electrical and Internet cooperatives that are common in rural areas of the U.S., or else to consolidate private networks and make them public. For example, my hometown of Toronto has one of the world's oldest municipal power utilities, Toronto Hydro, originally the Toronto Hydro-Electric System. By 1921, all of the various power companies that had sprung up to deliver electricity to the city from the generating stations at Niagara Falls had been purchased by the city and consolidated into a single larger network. It was more efficient and allowed for coordinated decision-making, and the incentive to profit from the network monopoly was replaced with a mandate to provide the service at the lowest cost to the users.

Collective Infrastructural Systems and Public Goods

If the resources delivered by infrastructural networks, like water or energy, aren't governed by the logic of the marketplace, neither are the networks themselves. Buying a car and "buying" the roads to drive on are completely different concepts because the economics of collective systems are fundamentally different from individually purchased consumer goods.

My car is a private good. The title is in my name—it belongs to me— and since I own it and use it, it's not available for others to use. Even when it's parked on a public street, I can keep other people from using it by locking the doors and taking the key with me. Physical objects lend themselves to being private goods, because atoms exist in discrete assemblages in time and space. Most of our personal belongings are private goods.

The roads that I drive on, on the other hand, are a public good. To an economist, that has a very specific and distinct meaning: in contrast to a private good, a public good is nonrivalrous and nonexcludable. "Nonrivalrous" means that one person having access to or enjoying a good does not preclude other people from doing the same. "Nonexcludable" means that people can't be prevented from using and benefiting from it. Public goods tend to be intangible. Even though the roads themselves are physical, what really matters is access to and use of it as a network, not the strip of asphalt itself. I can enjoy sitting in the Boston Common (established in 1634) on a warm summer day, and so can many other people. The city park is open to the sidewalk and the street on every side. Fences and gates around private parks are a way of physically enforcing exclusion, or at least signifying that they are excludable spaces.

Private goods have an obvious business model: make the thing, sell the thing. There's generally much less incentive for the private sector to produce nonexcludable goods, because if you can't prevent people from receiving a benefit and they no longer have a reason to pay for it, they presumably won't—the "free rider" problem. Too many free riders and too few paying customers means that the provider won't be able to make a profit if their business model is based on selling access to that benefit. With apologies to all my friends who work in digital media for the way I'm about to simplify a complex transition, I'm going to use journalism to illustrate.

For most of its existence, print journalism was able to limit the fraction of free riders and to fund reporting by selling a good that was *both* rivalrous and excludable, like a newspaper. I picture those scenes in old movies of a train full of businessmen on their morning commute, each reading his own copy of the morning paper. Because the articles and the information they contained were only available in print, the business model included selling the physical newspapers. The revenue from the purchase price (as well as from advertising) paid for the production and distribution of the papers themselves and for the journalism that filled their pages.

Intangibles, however, don't lend themselves to being private goods in the same way. Broadcast radio and television function almost exactly like streetlights: once those electromagnetic waves are released into the world, anyone in the service area who has a receiver can listen or watch. No one is excluded, and any number of people can listen, so it fits the economist's definition of a public good. That means that commercial broadcasters have always needed a business model for their news reporting that didn't depend on consumers paying for access, and it's typically been on-air advertising.

Digital communications upended media in part because of this shift from physical goods to intangible services. Any number of people can read the online version of an article simultaneously, but newspapers can still make their journalism excludable by putting it behind an online paywall. A subscription to the online version of *The New York Times* is what economists call a "club good"—it's excludable but nonrivalrous because, while only subscribers have access to the articles, anyone with access can read articles without diminishing the experience of others in any way. Online advertising is the analog to on-air commercials for broadcast and it remains, in all its increasingly creepy and intrusive forms, the default business model for "free" content on the Internet.

But noncommercial and publicly funded media, whether broadcast or online, turn the free rider problem on its head. If you believe that journalism and other programming have a value to their consumers and society at large that isn't about being as profitable as possible, then more people reading, listening, or watching is a free rider *win*, not a free rider problem. It means that, for a fixed investment of resources, more people can benefit. This is the logic that lies behind the public funding of radio and television all over the world, including systems like the British Broadcasting Corporation and the Canadian and Australian counterparts that it inspired.

This also gets us to why the economic definition and the everyday definition of "public good" are frequently confused or conflated. To an economist, the "good" in "public good" is in the sense of "goods and services," physical items that satisfy human needs or desires. Nonrivalrous and nonexcludable public goods often have large upfront costs but low or minimal incremental costs. For example, radio broadcasting requires the resources to build out a station and transmitter, and the funding to do reporting, produce programs, run a studio, and so forth. But once you've done all of this, it doesn't cost anything to have more listeners—they just need to turn on their radio (or, these days, to tune in online). But the flip side of this is that having more consumers isn't directly linked to more income or to reduced costs. For something to be a public good in the economics sense, therefore, there generally needs to be some impetus for making it available that goes beyond the profit motive. In the everyday sense of "public good," "good" suggests the opposite of harm, because in practice, public goods (in the economic sense) mostly provide broad, social benefits—that's why they exist at all.

If this is beginning to sound familiar, it's because it's also how many infrastructural systems behave. A network of roads is largely nonrivalrous—my taking the highway to work doesn't preclude other people from doing the same—and they are relatively nonexcludable. A municipal water system takes a lot of resources to get up and running—to deliver that first glassful of potable water—but once it's up, the incremental cost of delivering more water or adding households is quite small. A well-designed municipal system with sufficient capacity can be effectively nonrivalrous, where everyone gets the water they need.

Economic models aren't reality, of course, and there are few systems that are "pure" public goods. Roads and water supplies are rarely perfectly nonrivalrous. Traffic congestion slows travel on roads. In much of the world, drought can and often does limit the amount of available water, for household use and for agriculture. Parks get crowded. Few of these systems are also perfectly nonexcludable. A radio or a television broadcast still requires access to a radio or a TV. As individuals, our primary interface to infrastructural systems—electricity for artificial light, water for drinking and hygiene, fuel for cooking, and telecommunications—is where we live. Without a home, as in the significant unhoused communities in many major American cities, the ability to use and benefit from these infrastructure systems is limited or compromised. But the most common way of making infrastructural provision into an excludable good or service is simply through charging for its use—a water, electricity, heating, or telephone bill.

Services that do meet the strict economics definition of a public good often don't make business sense if the provider's primary motive is maximizing profit for its investors. But in many cases, it makes sense to build and maintain a system because it's in the best economic interest of its *users*. It can be financially worthwhile for them even (or especially) if there isn't a profit. This is how retail cooperatives work, for example—they increase the purchasing power of their members. It's not an accident that most of the systems that we consider to be infrastructure behave this way, because users benefit *individually* as a result of acting collectively with others. For example, about 85 percent of U.S. households have access to a public water supply and sewage treatment. (Because they're concentrated in cities, just 8 percent of water systems serve about 70 percent of the U.S. population.) The remaining households, which typically rely on wells for drinking water and septic systems for waste, almost always do so because there is no public system available to them, not because the private systems are considered to be a superior alternative. Network effects mean that universal provision has the potential to be both cheaper and better for *each* of its users; the fact that everyone has access to it is almost a side effect. This also helps explain why the provision of other basic human needs, like food and housing, isn't considered to be infrastructure: if your neighbors aren't adequately housed, it doesn't negatively affect your household in the same way as it does if they don't have adequate sanitation. And unlike your electricity or telecommunications provision, your groceries generally don't get cheaper if your neighbors eat too. Even though food and housing security for all is a public good, it's "only" in the sense of having social value, not in the economic sense. I found this to be something of a grim realization because of how far it went toward explaining how Americans can take universal water provision, electricity, and roads mostly for granted while ensuring that everyone in their community is fed and housed remains an ongoing political and social challenge.

Compared to the familiar physical objects of our world, then, which are rivalrous and excludable by default, networks and systems can behave in unusual or even counterintuitive ways. That makes calculating their cost, benefit, and value less straightforward, not least because they can enable other kinds of benefits that are less easily captured on a balance sheet. These systems are often public, because building and maintaining them is definitionally in the public interest. Public control and operation allow for expanded access and unified networks while sidestepping the risks to consumers of network monopolies in private hands. Electricity, water, and transportation systems worldwide all followed this pattern as well, merging fragmented private networks into unified public systems. This includes public transit in cities such as Boston, New York, and London—it made a lot more sense that the Underground map is a tangled mess once I learned that a patchwork of private companies was providing subway services, largely as a way to avoid the congested surface roads, before they were brought under public ownership and consolidated into the Transport for London network.

If infrastructural systems substantially reduce the cost of provision for users, and they behave very much like public goods in the strict economic sense, and they provide basic necessities or survival needs, then it becomes clear why the term "public good" is confusing. Provision of these services is a public good in the *moral* sense, as in the opposite of evil. The imperative to offer aid to others, including strangers—to provide food, water, and shelter—is part of religions and cultures the world over. Especially when infrastructural systems provide for survival needs, like water or heat, their provision intersects with these social values. That moral element is thrown into especially sharp relief when services are shut off, because usually the presence of a network means that the resources it provides will be available in any case. Not allowing access because someone is unable to pay their bills is then actively *withholding* basic human necessities, rather than neglect or lack of capacity. It's why denying water to those who needed it was a way of conveying, in just a few moments of screen time, that Immortan Joe was monstrous.

Beyond the shared and moral benefits of access, infrastructural systems offer benefits that are so integral to the way we live our lives that they are hard to even value. In economics terms, these are externalities—benefits (or costs) that are not fully accounted for in the pricing of a transaction.

Infrastructure as Externalities

The best honey I've ever tasted comes from my friend John Pittman's bees, which live in hives in the backyard of his home in New Hampshire. On a sunny summer afternoon, they leave and return to the hive in a steady stream, a softly buzzing flight path that reminds me of how closely spaced planes take off and land at busy airports during peak times. The honeybees collect nectar from the wildflowers and fruit trees that grow on neighboring yards, bring it back to their hive, and turn it into extravagantly floral-scented golden sweetness. From a strictly economic point of view, his neighbors could argue that John is getting positive benefits from their properties without paying for the privilege of accessing them. Of course, it's also possible that they're the ones who benefit from the bees, which pollinate flowers as they collect nectar. Most honeybees in the U.S., in fact, get driven on trucks across the country to fields and orchards, following the seasonal blossoms, on a "pay for pollination" basis. John's neighbors may be making a fruitful (ahem) exchange which they are simply unaware of.

Externalities, to economists, are values or costs that aren't accounted for by the market, because they accrue to someone who isn't part of the transaction and therefore often has no choice in whether it happens or not. If a company pollutes the watershed where a facility is located, but doesn't pay to either remediate the water source or to cover the costs associated with the impacts on local residents, it's a *negative* externality that affects people other than the customers who paid the company. From secondhand smoke to "forever" chemicals, this particular externality is ubiquitous and is the major driver for environmental regulation, which typically accounts for the costs by requiring companies either to mitigate the impact (like installing scrubbers to remove the acid rain-causing sulfur oxides from coal-burning power plants) or to remediate the effects. The costs for these interventions then show up on balance sheets and profit forecasts.

Conversely, a positive externality is when good things happen to people who haven't paid for them directly, like when John's honeybees collect nectar from his neighbor's land. In much the same way that Schrödinger's cat is both alive and dead until someone looks in the box and observes which, the cost of something only becomes real the moment a buyer trades money for it. Positive externalities are hard to value in dollar terms *because* they're not paid for.

When we try to determine the economic value of collective systems that are so important and ubiquitous that we build other systems on top of them, the dollar estimates can quickly become astronomically high, almost impossible to conceptualize. The 2021 budget of the U.S. National Weather

Service (NWS) was about $1.2 billion dollars. The commercial weather business in the country, most of which is built on the data provided by the NWS, is a *seven* billion dollar industry, and the direct economic benefit of weather forecasting is estimated at $13 billion. Location services are an even more striking example. It costs the U.S. government about a billion dollars a year to maintain GPS satellites and systems. In 2016, Greg Milner estimated the direct value of the global GPS market to be around $30 billion, having tripled from five years earlier. But when Milner asked for an estimate of the dollar value of the *entire* GPS market—the chips that are in receivers, the products and services built on top of it—an industry expert declined to answer, saying only that it was in the trillions, "meaningless for anyone but a scholar." As a scholar, I can say it *does* carry meaning for me, not as data but as a signal, a flashing yellow light—if the dollar estimates seem laughably, incomprehensibly high, it means that using money as a metric is, if not precisely *wrong*, at least desperately incomplete. Weather forecasting or satellite mapping, as services that cost a lot to provide but which then have very low incremental costs to access and use, are close to being true public goods even in the pure economic sense, since they're nearly nonrivalrous. The more they're used, the more valuable they are, in ways that go far beyond the economic value of the industries built on this data. Like the industry expert, I agree that the true value of these systems is literally incalculable, but it's because they enable systems and behaviors that wouldn't be possible without them, and they help us to protect the irreplaceable. Collective infrastructural systems, especially public utilities, are giant masses of positive externalities. Location services help us navigate with confidence no matter where we are or what transportation mode we're using, to meet up with family and friends, to track objects from a teddy bear to a freight train, and to go on adventures. Weather forecasting helps us decide if we want to go for a walk or if we need to evacuate a major city. And these aren't even the critical utilities, like clean water, electricity, or transportation. How could you even begin to put a dollar value on the benefits that access to electricity brings to our lives?

Infrastructure as Economic Driver, and More

It's a staple of news reporting around major sporting events like the Olympics or the World Cup—a spate of articles about the "lost productivity" that results from people watching or discussing sports during their working hours. It's a familiar example of how we've gotten accustomed to putting everything in dollar terms. While economic analysis is undeniably useful, it's also woefully inadequate, especially when we're talking about infrastructural systems. In the same way that most calculations of lost productivity due to sports fail to account for any positive effects, even ones that might directly benefit the employer, like employee cohesion and morale, any discussion of monetary costs and benefits only captures a small part of the value of what infrastructure is for.

The value of infrastructure is often framed as a form of cooperation that underpins the economy. Infrastructural systems make it possible for individuals to participate in the workforce and to be economic actors, contributing to the economy and creating wealth. The universal provision of services and facilities (like electricity or roads) means that they are both better and cheaper than what individuals or smaller groups could provide for themselves, and the time and labor that's freed up can be used in support of industry. This kind of argument was made to taxpaying residents of 1890s Boston: by funding municipal improvements like water mains, sewage treatment, and graded roads, more people could be gainfully employed, to the benefit of the Commonwealth of Massachusetts. Because much of the survival labor they replace (like obtaining clean water or fuel for cooking) is highly gendered, the role of these systems in underpinning agency, including economic agency, can be especially profound for women. I think it's probably not a coincidence that the women's suffrage movements in the UK, U.S., and Canada all grew alongside water, sewage, utility gas, and electrical systems in the late 1800s and early 1900s, a distribution of power in more ways than one.

Freeing up the labor of workers is only the most basic economic utility of infrastructure. Private companies also benefit directly from infrastructural systems, and this has long been the primary economic argument for public investment—that it underpins economic development and growth. Corporations take advantage of transportation systems and shipping to move goods around, along with postal mail and telephony and its successors to coordinate their businesses. They can sell washing machines because water and electricity are utilities, and cars because there are public roads to drive them on. Companies sell computers (and tablets, and phones, and online advertising) because communities enjoy widespread Internet access. Infrastructural systems, public or not, also take advantage of public resources, from hydroelectric dams built on rivers to the limited electromagnetic spectrum that's parceled out for telecommunications providers. How do we figure out the benefit to corporations of the existence of infrastructural systems? How do we figure out the value to a company like Amazon, and by extension its shareholders, of not just the United States Postal Service last-mile delivery service that they might pay for directly, but all the public roads and municipal transit for employees, GPS services, weather forecasting, transportation for workers— all the services that are necessary for businesses to function and for customers to purchase their products? The economic benefits might be calculable— barely—for services like GPS and weather forecasting, but they are almost impossible to encompass for roads or electricity or even clean water.

When a company puts something in the world that someone benefits from without paying for directly, it's a "positive externality." But when corporations and their shareholders benefit from common-pool resources, everything from the natural environment to infrastructural systems built with public money, that's often just considered to be the natural order of things—it's what they're for. There almost isn't a word for it, besides phrases like "public investment for economic development." But there's an implicit deal here, which is that at least some of the economic benefits that these systems make possible will be reinvested into these systems to keep them operational. Without corporate taxes that adequately reflect the enormous subsidies that companies receive in the form of public infrastructural systems, public investment in infrastructure becomes a massive transfer of wealth from individual taxpayers to corporations and to their investors and shareholders. This is extractive capitalism, where wealth is extracted from workers, communities, and common-pool resources and the negative impacts and externalities are left unaddressed.

This matters for a lot of reasons, of course, but the one I want to focus on for a moment is that economic agency is just one slice, in Amartya Sen's formulation, of living the kinds of lives we have reason to value. Sen describes money as a general-purpose means of providing freedom. The difference between money and collective systems like infrastructural provision is that while both provide agency and freedom to individuals, money does it at the level of individuals, in an atomized way. Infrastructural networks, by their nature, increase individual freedoms *collectively.*

It's been argued that the Great Enrichment invokes a virtuous circle: an expansion of social rights led to more peer-to-peer cooperation between individuals and hence more economic surplus, which in turn leads to more social rights. A similar argument holds for networked systems, particularly public infrastructure with universal provision: any decision to build them out in the first place presupposes some social rights, but they also underpin the continuing development and recognition of such rights. In whichever direction the causality lies, there's a relationship between increased social, civic, and economic participation and the universal provision of infrastructural services, which is why we associate these infrastructural systems not just with economic development but also with equity and increasing enfranchisement. What that entails, precisely, changes with time and place. Sen, writing about people and communities with little access to either wealth or infrastructure, summed it up:

> Economic growth cannot sensibly be treated as an end in itself. Development has to be more concerned with enhancing the lives we lead and the freedoms we enjoy. Expanding the freedoms that we have reason to value not only makes our lives richer and more unfettered, but also allows us to be fuller social persons, exercising our own volitions and interacting with—and influencing—the world in which we live.

We can observe this expansion of freedoms over the course of the twentieth century, and into the twenty-first. Over the course of my adult life, I've had the opportunity to observe this expansion with new infrastructural systems firsthand.

The Infrastructural Pathway

During the time I lived in the city, Shoreditch—a hip neighborhood in East London, filled with bars, coffee shops, restaurants, and tech companies—was home to a venue that took its name from the letters spelled out in lightbulbs over the doors: ELECTRICITY SHOWROOM. A few years later, I was in a cab in Mumbai, India, heading up the Colaba Causeway, and I spotted a period building whose original signage read ELECTRIC HOUSE. I knew exactly what I was looking at and why it was there, halfway around the world.

When electricity first began to be commercially available for domestic use, it was largely as a luxury in the homes of the wealthy. A typical early setup would have involved the installation of a small generator along with the wiring necessary to replace gas lighting with incandescent lightbulbs, which were cleaner, brighter, safer, and which required less labor to operate and maintain. As electricity began to be more widely available, initially from coal-fired generating stations in urban areas, network effects gave local providers an incentive to grow and densify their customer base. More users wired into their circuits lowered the per-customer cost of provision and cemented their network monopoly. Providers needed to persuade local households of the value of residential electricity, so the Electricity Showroom in Shoreditch and the Electric House in Mumbai were built to display the wide variety of novel labor-saving electrical appliances, like clothes irons and toasters. They were the electricity equivalent of the storefronts of wireless phone service providers with their displays of mobile handsets, showcasing the latest high-tech devices not so much for their own sake but as a means of convincing customers to sign up for the service.

Showrooms like these were initially part of the expansion of electricity provision from a luxury into a service which many more people could afford. The Electric House in Mumbai was built as the showroom of a private concern, the Bombay Electric Supply and Tramway Company (the seed of what would become the city's municipal electrical utility and public transport system). But the one in Shoreditch was built by the local council to hasten the adoption of electric lighting by making it easy to sign up for electricity: the council would send someone out to install the wiring and lamps for you for free, and you could even pick out a lampshade that matched your sofa. For households, there was the value proposition of more and brighter illumination, but there was also a positive, significant, nonmonetary externality: electric lighting meant a reduced risk of fires, which was especially important in the poorer, densely built and occupied areas of East London. Widespread electricity provision was understood to be a social good. One major reason why the early private electric companies were taken over by cities to become municipal providers was to transform the risks of privately held network monopolies into a self-propelled virtuous circle, where reduced costs of provision could be passed on to customers as more and more buildings and households were wired into the electrical grid. Electricity went from being a service (in the market sense of "goods and services," something you could pay for or not, depending on how much you valued it) to a utility. One way of thinking about a utility, especially a public utility, is that it's not just a service that's valuable in its own right. It's one that's so necessary that, by default, everyone has access to it, all of the time. Utilities provide the foundational layers on top of which additional systems are built, including the goods and services that take advantage of them.

Having recognized this directional pattern of niche or luxury, to service, to utility, we can observe it elsewhere. Once upon a time, GPS receivers were a niche or a luxury, the province of boaters and adventurers. Then these kinds of real-time location services found their way into personal vehicles as a nice-to-have service, like in a rental car in an unfamiliar city. But once locational information became widespread, as on mobile phones, systems could be built that used it continuously—not just mapping or infrastructural services like delivery tracking or transportation, but everything from hyperlocal real-time weather forecasting to social services that enable users to find nearby friends (and not-yet-friends) to location-based games like Pokémon Go to location tracking for ad targeting or other, less benign, forms of surveillance.

Having become a utility, where its presence is taken for granted as a substrate layer for other systems, the next step is to become a political right. That might be because the system meets a basic survival need, like potable water, or because it enables agency, as artificial light and access to energy generally, or it might be directly political, in the way that civic systems, including public education, access to social services, and democratic participation, are increasingly or even exclusively accessed online. As infrastructural systems progress from utilities into political rights, we are effectively recognizing and codifying new forms of agency, or at least ways of providing that agency; in Sen's terms, it's an expansion of freedoms that we have reason to value. This transition occurs, in part, because the spread of these new systems and the new forms of agency that they allow is a double-edged sword. Since they make resources cheaper and easier to access, it becomes increasingly difficult to opt out in favor of alternatives; in Neil Gaiman's succinct phrasing, tools can be the subtlest of traps. What's more, as a system becomes widespread, lack of access to it, for whatever reason, becomes increasingly disenfranchising because its presence starts to be just taken for granted and alternatives become less and less available. It's no longer about your individual decisions: if access is assumed, then *not* having access limits your ability to fully participate in society. It's analogous to the commitment to public education and the rise of universal literacy over the past two centuries, from a 15 percent global literacy rate in 1800 to about 85 percent now. Part of why utilities become political rights is precisely in response to (or to avoid) that inadvertent disenfranchisement.

The transition from a utility to a political right is a general progression, not a clean, stepwise process. It results from the commitment and work of political activists. Access to sufficient clean water, regardless of whether it's provided through collective provision or not, is explicitly enshrined in the constitution of South Africa as an explicit political right. Political rights to infrastructural systems are often expressed in the form of policy, as with universal service mandates that guarantee access to communities that might otherwise be underserved, including low-income households or rural areas which fall on the wrong side of network effects. In the twentieth century, that meant access to grid electricity and to telephones. In the twenty-first century, countries have begun to put universal service mandates for broadband access into place. Finland was the first, in 2009, and Canada mandated it in 2016. There's a paradox here that helps explain why these countries were out of the gate early: the bigger or the more thinly populated a country is, the more of a challenge it is to build out broadband networks, but also the more valuable it's likely to be to its residents. The peak of the global COVID-19 pandemic in 2020 and 2021, with the associated movement of many public services online, is likely to be a driver for more such mandates.

Finally, access to infrastructural networks is often *implicitly* a political right. After Hurricane Maria, when parts of Puerto Rico lost electricity for as long as a year, I often heard people say something like "these are *Americans*."

Since then, I've heard it again and again after different infrastructural failures and crises. Stressing that point is a clear indication that access to public utilities like electricity and clean water is seen as a political right, because if it weren't, nationality would be irrelevant.

The Dinorwig Power Station is among my very favorite infrastructural installations not just because of the eyewatering scale—Electric Mountain!— or the way that it uses natural features in the landscape to elegantly accomplish a goal. Dinorwig tugs at my heart, not just my head, because of how it incorporates other values besides efficiency. It was built out as the local slate quarrying industry was in decline, and the workers on the project included retrained former miners. When Marchlyn Mawr and Llyn Peris were turned into reservoirs, the arctic char living in the lakes were relocated. (This seemed like a lot of effort to go to for some fish—arctic char are widespread and commercially fished across northern Canada—until I learned that they are locally endangered in the UK, just small populations that had managed to hang on in a handful of Welsh lakes since the end of the last ice age and which probably wouldn't survive their homes being regularly drained and refilled.) The entire power station and its facilities were designed to blend into the scenic landscape. There are no marching rows of electricity pylons, since the transmission lines are buried underground.

Infrastructural systems, by their nature, are more than just technical— they're inextricably social and political because they are intrinsically collective. This is especially true for networks, which are shaped by the sustained relationships of the people who live in the places they connect and, conversely, the infrastructural systems that are embedded in the landscape form part of that sustained relationship. Because they incorporate nonmonetary externalities, both positive and negative, they can't easily be valued or assessed like a consumer good, where it's "worth it" to buy something or not. So they don't lend themselves to decision-making that focuses solely on the costs or the returns on investment. An infrastructural network can encode and promulgate a set of values: *everyone should have access to clean water,* or *electricity is a necessity,* or *personal mobility is a human right,* or *a healthy population is important,* or *broadband access is required to fully participate in civic society,* or even *endangered fish should be protected.* If infrastructural systems meet basic human needs, providing agency and freedom, the form they take depends on cultural norms and expectations; in turn, the systems set and define those norms and expectations. Pumped-storage hydroelectric stations are used globally but I love Dinorwig in particular because it really foregrounds how infrastructural systems are built to fulfill human desires, and how specific, contingent, and culturally situated these desires can be. What do people want? How can these desires be fulfilled? How are the desires of different communities balanced? How are the necessary resources distributed? These are political questions, not engineering questions. But in infrastructure, the political and the engineering questions are inextricably linked.

The most radically important promise of infrastructural systems is not just that they underpin our individual agency but that they can do so in a way that is democratic and universal. Most technologies are sold on the premise of being liberatory, but infrastructural systems have genuinely demonstrated that they are, and they have at least the potential to provide that freedom in an equitable way. Over and over again, as new technological systems have been invented and the networks have begun to be built out, their efficiency, their value, and the way they enable personal, civic, and economic agency have been recognized, along with technical, social, and economic incentives to move toward universal provision—to treat the services they provide as a public good, in all senses of the phrase. It's why good infrastructure is strongly correlated with not only access to resources, but with equitable, stable political systems. As infrastructural networks become more widespread and reliable, we begin to build other systems on top of them; this provides a powerful incentive to make them even more widespread, reliable, and enduring. Around the world they have been both the roots and the fruits of sustained economic and political cooperation:

enabling groups to use and steward resources that are held in common for collective benefit, and to make decisions that are in the best interests of the community. Shared systems are used to meet basic needs, to provide services, to foster social connections, and to provide access to physical goods. Infrastructural systems are how we take care of each other and plan for a future of long-term investment and cooperation.

For a long, long time, this is how I understood infrastructural systems to work. And all of this is true. But slowly I began to realize that I had only understood half of the story.

Six

THE POLITICAL CONTEXT
OF INFRASTRUCTURE

Growing up I had the chance to see Niagara Falls in every season because, like many South Asian immigrants living in Toronto, it's where we went whenever friends and family came to visit. In the teeth of a cold snap in the depths of winter, Niagara Falls are an icy wonderland. Every surface is wreathed in frost from the spray and the river is a mass of ice. But even in the grip of the deepest, longest freeze, the water never stops flowing. The cataracts remain an awe-inspiring sight for visitors from around the world, every day of the year. Niagara Falls might be a natural wonder, but they're shaped by paired hydroelectric power stations and an international treaty.

A few miles upstream of the Falls, next to the scenic parkway that runs alongside the Niagara River, are two large, windowless buildings. They house the controls and equipment for the intake valves that divert water from the river into huge tunnels that carry it around Niagara Falls and its rainbows, whirlpool, and sightseeing boats, and under the tourist-filled city center, to be delivered a few miles further downstream where the two hydroelectric generating stations face each other across the Niagara River and so across a national border. Despite being in different countries, the power plants are part of the same electrical grid, coordinating energy usage across much of the eastern half of North America. Both utilities have agreed to limit the amount of water they divert around Niagara Falls, and therefore the amount of power they can produce, to ensure there is always at least a curtain of water flowing over the edge for visitors to appreciate. That's especially important on hot summer days when the river is low, energy demand is high, and it's high season for tourists, as on the day I visited the New York Power Authority's hydroelectric power station. I found it surprisingly welcoming and fun, although I'm pretty sure it's mostly geared toward schoolchildren, with playful interactive exhibits on how the electrical grid works, models and displays about hydroelectric power generation, and even a movie theater equipped with vibrating seats, flashing lights, and sprays of water, all to create an immersive experience as we learn about how flowing droplets of water are transformed into zooming electrons. From the plant's observation deck, I stood and looked out over the Niagara Gorge, across the turquoise-blue river to the Ontario Power Generation facility on the other side and downstream to a bridge that serves as a border crossing. As someone with close ties to both countries, this physical evidence of careful, ongoing international cooperation means a lot to me.

In the visitor center, historical exhibits tell the story of how, in 1956, the hydroelectric generating station at the site collapsed in a landslide caused by water seeping into the riverbank behind it. Robert Moses, the urban planner best known for his role in reshaping the roads and bridges of New York City, had recently become the first chairman of the New York Power Authority. Moses took charge of the Niagara Power Project and rebuilt the plant, now named after him, as a secure power source for his city. The project was approved in 1957 and opened just four years later in 1961.

It can't be seen from the visitor center, but the design of the plant incorporates a reservoir. On maps and in aerial photos, it's visible as a polygonal lake behind the plant, away from the river, and it serves the same purpose as the high mountain lake at Dinorwig, storing water that can be used to generate electricity during times of high demand and then refilled when demand is low. On the other side of the ruler-straight eastern edge of the reservoir are Tuscarora Nation Reserve lands. Having decided on the land they wanted for the reservoir, Moses and the Federal Power Commission went to the U.S. Supreme Court for the right to break a treaty obligation and seize the site for the reservoir by eminent domain. In 1960, the ruling came down in their favor. In the dissenting opinion, Justice Hugo Black wrote, "Some things are worth more than money and the costs of a new enterprise . . . I regret that this Court is to be the governmental agency that breaks faith with this dependent people. Great nations, like great men, should keep their word."

Who Are These Systems For?

The nature of people acting collectively is that they form, well, a collective. Collective norms around infrastructural systems can manifest as universal provision, a commitment to provide an acceptable standard of service to everyone. South Africa, for example, which has access to water in its constitution as a basic human right, operationalizes it with specific provisions, including a free monthly allowance for every household. If infrastructural systems are a physical manifestation of social cooperation, that also means they're a physical manifestation of the values and norms for the group. So as part of the transition from a service to a utility, this idea of what's "normal" also undergoes a transition.

Flush toilets had already begun to catch on in London in the middle of the nineteenth century, so when Joseph Bazalgette designed and oversaw the construction of London's sewage system, he made the decision to base it on the strategy of using water to dilute waste and rinse it into sewers, to flow downhill and downstream, into the estuary of the Thames River to be washed out to sea. Wastewater is now usually treated before being released, and this system of flush toilets, sewers, centralized treatment, and outflow into a body of water is considered the "gold standard" for managing sanitary sewage. It's efficient in the way it leverages flow through a network, but it also encodes a set of social norms which were initially local but which are now global. To begin with, Bazalgette's approach made perfect sense in densely populated and rainy London, but it makes much less sense in places where water isn't so plentiful, which is a large fraction of the planet. Remember how surprised I was at my local wastewater treatment plant that the sewage was clear and nearly colorless? The way we deal with our smelly, disgust-inducing, pathogen-laden waste is by contaminating large quantities of otherwise clean, potable water. But there's a second issue, which is that discharging the effluent into waterways means that our food systems and sewer systems can together form a one-way trip for the nutrients in soil: from the ground, into food, into us, and out into the water through our sewers. In industrial agriculture, those nutrients are mostly replaced with nitrate fertilizers derived from petroleum using the Haber process, which enables food production at scale but which is also a significant contributor to global greenhouse gas emissions. Other types of systems for managing waste, like individual or community-scale septic systems, are often seen as stepping stones to sewer lines and waste treatment. Composting toilets, in which waste is converted to fertilizer which can then be returned to the soil, have been generally regarded as the province of weird hippies in North America, and are seen as subpar, second-class systems by much of the rest of the world.

If we say that infrastructural systems are built in response to social norms and also shape social norms, that leads to an obvious question: Whose norms? What is the "social" in question? The technologies and systems that are available shape expectations: not just the way that we live our lives, but the way that we believe that lives should be led. Across a community, this is the domain of politics. As new technologies become available and are adopted, they enter this realm, leading to policy like the Rural Electrification Administration in the U.S. (and its counterparts worldwide): the cultural norm that everyone should have access to electricity, no matter where they live. Infrastructural systems have been an important means of empowering groups to have full civic and economic participation, so they're not just the result of politics—they embody and enable politics. That also means that they're not just about benign or nearly benign norms about tea and flush toilets. They are manifestations of power, of structural inequities, and of systemic biases around who deserves access and agency, and who does not.

The group that controls the social norms around building infrastructural systems is very likely to be the group with the power to make the decision about what systems get built and where, about who benefits from them, and who bears the costs. Dinorwig is striking for the care that was taken to mitigate its negative impact. It's a sharp contrast to the impact of the New York Power Authority plant at Niagara Falls, where the construction had already been approved and begun before the Supreme Court decided the case. Robert Moses expected to win the legal challenges over Tuscarora Nation land. He had good reason to: across the U.S. and Canada, dams and reservoirs for hydroelectricity had displaced Indigenous communities before and would continue to do so after Niagara. The cultural history and norms of those peoples were considered secondary because norms are set and decisions are made by those who hold the power to do so. In this light, infrastructural systems can be seen, above all, as physical manifestations of the same kinds of cooperation that led to the Great Enrichment. They are networks that enable an exponential growth of wealth, power, and agency, made possible by energy use and resource extraction. Seeing them this way leads to obvious questions: Wealth, power, and agency for whom? Energy and resources from where?

Recontextualizing Mapping and Networks

On August 14, 1947, when my father was twelve, he woke up in Pakistan. The day before, his family, his home, and his bed had all been in India, but at midnight, Pakistan had become independent both from Britain and from the rest of India, which gained its own independence twenty-four hours later. With the end of British rule, the two countries were divided along what became known as the Radcliffe Line. The Viceroy of India, Lord Mountbatten, had tapped Sir Cyril Radcliffe, a lawyer without any training in geography or cartography and who had, infamously, never been to India before, to lead the team to draw up the new borders. Deciding on where to place the boundaries would have been a complex and thankless effort even without the backdrop of open sectarian conflict, much less the impossibly tight deadline: Radcliffe spent just five weeks on a task whose consequences have echoed down the decades. Partition, as it's known, became one of the largest migrations in human history. Millions fled communities convulsed with ethnic and religious violence taking what belongings they could carry, preyed on by opportunistic armed gangs. My father and his family were among them, leaving their home a few months after Partition and living in a refugee camp for months before eventually settling in Bhopal.

In the nineteenth century, mapping was a significant scientific preoccupation and a source of national pride. The sustained, even heroic, efforts to describe the exact geometry of our planet which were undertaken at that time continue to underpin all modern maps, including satellite systems. But you can almost infer the imperatives for mapping in India—scientific, sure, but mostly commercial and military—just from the dates. The Great Trigonometrical Survey of India was started in 1802, when the East India Company had taken over the entire subcontinent, and just a decade after the triangulation of Great Britain itself. In 1858, the East India Company was nationalized, bringing India fully into the British Empire, and then in 1947, those maps were used to define the boundaries of the newly independent states. In the century and a half in between, the maps were used to make the subcontinent legible to its imperial overseers, enabling Britain to cement its lucrative control, taxation, and development of the subcontinent. Mapping also served as the underlayer to the infrastructural networks built to facilitate the movement of resources around India, but especially *out* of it. The Suez Canal, which opened in 1869 with primarily French investment and under French control, was taken over by the British in 1882, first by buying out the local investor and then by invading the region in order to impose military control. Control of the Suez made it easier—faster and thus cheaper—to ship goods and resources from India and other eastern colonies for the benefit of the people living in the UK. The new submarine and overland telegraph cables that were laid by 1870 meant that political control of India from Britain could be tighter than ever before. Even the passenger rail network in India, often considered a positive legacy of the British Raj, was largely built to facilitate the movement of goods and of British soldiers. It was paid for by guaranteeing private investors in the UK a 5 percent return on their investment, making it a net transfer of wealth from the people of the subcontinent to middle-class investors in the UK.

During the half a year or so that I lived in London, I regularly walked past India House, built after World War I to house the administrative offices for the subcontinent as part of the political devolution of India in advance of the region gaining independence (it now houses the Indian High Commission). The building is decorated with emblems representing each of the provinces, including an elephant for New Delhi and a moon over a gateway to represent the North-West Frontier Province, where my parents were born. But the real monument to British control of India was a block or so south, on the embankment of the Thames River. It was under my feet, and it was in the railway stations, lines, and bridges all around me. Not for nothing does the golden age of Victorian public works coincide with the peak of the British Empire. Over the course of the Raj, the infrastructural systems—first mapping, then railways, canals, and early telecommunications—that were built out across India facilitated political control and extraction of resources. These new technologies were powered by new sources of energy, like coal, and much of the new wealth that was produced was sent to the UK. Historians have estimated India's share of the world's economy to have been 25 percent before British control, dropping to just 4 percent or so by the time the country gained its independence. Meanwhile, of course, the wealth flowing into the metropole was available to be invested in collective infrastructural projects for its own benefit, like Bazalgette's sewer system and the cleaner river that was the result.

Every story of infrastructure, trade, and European colonialism is uniquely complex, and I've just sketched the smallest part of one of those histories here. I'm not going to commit the fallacy of thinking that human societies can be explained by scientific and engineering principles. But framing this history in terms of exponential growth—investment, generation of output, reinvestment of that output to generate even more output—is a useful way of thinking about cooperative behavior in societies in general and infrastructural systems in particular. Investment into these collective systems can yield outsized returns, both monetary and nonmonetary. Growing up in Canada meant that I had an infrastructural birthright, the beneficiary of two centuries of unprecedented social, economic, and technological cooperation, with the snowballing wealth, power (in both the political and the engineering sense), and agency that resulted from it. That birthright set me up for my life of continuing freedom and security, including being positioned to benefit from new infrastructural systems as they were developed and made available, like GPS and the Internet.

While collective infrastructural systems can be used to benefit everyone in a community, they are also used to funnel resources away from a larger area to that community. I'm focusing on India because it ties into my family's history, but the sun never set on the British Empire. The "Scramble for Africa" in the decades around the turn of the twentieth century, during which direct European control of the continent grew from about 10 to about 90 percent, was largely made possible by new technologies of communication and transportation, like the coal-powered steamship. When I walk around central London, all of the Victorian-era civil engineering around me serves as an inescapable physical reminder of the bodies and labor of people around the world, including colonized and enslaved peoples. Infrastructural systems are designed to efficiently concentrate energy and resources to particular people in particular places. More money, more energy, more technological complexity, and more cooperation mean that these networks can get bigger and bigger, transporting resources for the benefit of whoever controls the networks, which means they get more power and wealth and they can continue to grow out these systems, and to build new systems on top of them. This isn't a side effect of infrastructure. It's what they are.

Centralize the Benefits, Displace the Harms

In my home, clean water comes out of a tap, and sewage disappears into a drain. Parcels arrive in my mailbox, and I put the waste packaging in the recycling or the garbage to be taken far away. The electricity I use comes from generating plants on an electrical grid that spans a good chunk of North America (and no longer from coal-fired plants in my neighborhood—I can still see where several of them were once located, even if the buildings have since been repurposed). The natural gas that heats my apartment and hot water and which I use to cook my dinner is extracted thousands of miles away. It's burned to water and carbon dioxide, and those combustion products don't stay in my home—they're vented into the atmosphere, along with the gases from my car's exhaust pipe and the jet engines from planes when I travel.

The infrastructure I rely on requires energy to function and is built out on physical networks. All of that energy and all of those materials have to come from somewhere. Without careful mitigation, every stage of this process—extraction, fabrication, transportation, consumption, and use—has negative effects: pollution and waste, noise and congestion, and more. All of those harms end up somewhere. There might be environmental degradation associated with the extraction of raw materials, like mining sites. There might be toxic by-products of chemical processing. Thermal electricity generation produces a range of waste products, including the air pollution from coal- and gas-fired electrical generating stations and the solid radioactive waste from nuclear power plants. There's noise and pollution associated with transportation, largely produced by vehicles themselves. There's solid waste, whatever is left over when physical goods reach the end of their useful life, from microplastics to e-waste. There's the effluent from a sewage plant or septic systems, or residual brine from desalinating seawater for drinking. How are these harms managed?

The possible answers to that question are closely linked to others: Who are infrastructural systems *for*? Who benefits, who pays the costs, and where are the resources coming from? As Langdon Winner put it, all artifacts have politics: they embody and encode certain social structures, ways of doing things, and what interests are served. That's literally true in the case of infrastructural systems. A federal interstate highway system that makes it easy to move goods around the country says something about how commerce is valued; a modern and well-maintained passenger rail network says something about human mobility as a policy priority. If a water treatment plant embodies politics, so too does the storage reservoir at a hydroelectric station that displaces an Indigenous community.

One of the reasons why I appreciate the Dinorwig Power Station so much is because of the decisions made to minimize harms, like rehoming the local fish or burying the transmission lines so as not to spoil the view. What this reveals is that the beneficiaries of Electric Mountain and the community that would be negatively affected were considered to be largely the same people: the British public, plugging their kettles in and spending bank holiday weekends at Snowdonia National Park. This overlap provides an incentive to mitigate the negative effects of the project, just like there is for the cooperation between Ontario and New York State that ensures enough water flows over Niagara Falls to preserve its appeal to tourists on both sides of the border.

This kind of alignment is rare. In Boston, the rail lines that head southwest out of the city were originally intended to be sunk below grade but in an open trench, a much cheaper approach than covering it over to form a tunnel. Activists from the predominantly African American neighborhoods along the tracks came to planning meetings for the Southwest Corridor and interrupted them by playing recordings at a volume that matched the expected decibel levels of uncovered trains, thereby demonstrating exactly how disruptive the noise would be. They got the point across, and the tracks now run in a fully underground tunnel. There are literally thousands of oil derricks across Los Angeles, including both orphaned wells and active sites (which were finally banned in January 2022) but the degree to which any potential health impacts have been mitigated depends on where they are and who's most affected. These are just two examples of how all decisions about location and mitigation are intrinsically political and can be used to systematically advantage or disadvantage groups, even within a single city.

The more resources that are consumed, the higher the cost that needs to be paid. Wealth makes it possible to pay people and communities to deal with the harms; conversely, the less wealth that the recipients have, the more likely they are to accept the terms. At the national level, coal mining or natural gas extraction in Pennsylvania, Ohio, and West Virginia powers electrical generation all over the U.S., but the costs of that extraction are locally borne. And just as many of the benefits of infrastructure are in the form of nonmonetary positive externalities, the costs are largely in the form of nonmonetary negative externalities, hard to account for. How do you put a dollar value on mountaintop removal mining? Or polluting a region's water sources? And who decides what that value is? For whose benefit? Wealth might make it possible to pay people and communities to accept harms; conversely, the less wealth that the recipients have, the more likely they are to accept the terms. But that doesn't mean the terms are fair. Worse, it means that existing inequalities are compounded. If infrastructure is a way to provide the comfort, safety, and freedom from basic needs that underpin agency, including economic agency, then these sorts of harms do the opposite, further eroding access to opportunity of all kinds.

Infrastructural systems that make it easy to bring resources and benefits to a group almost always make it just as easy and efficient to displace the harms associated with extraction, consumption, and waste onto *different* groups of people, whether nearby or on the other side of the world. As these infrastructural systems have grown in size, population reach, and geographic extent, the magnitude of harms—the negative impacts associated with vastly increased usage of energy and consumption of raw materials— has grown commensurately. It's become possible to displace the harms of these systems farther and farther away, creating a severe moral hazard: any incentive to mitigate harms is reduced by the ability to retain the benefits while offloading the costs onto, bluntly, someone else. Or, in the case of pollutants in the oceans or the atmosphere, *everyone* else. Whether you see networked infrastructure as collective systems that serve the common good or as systems that are designed to extract resources and deliver them to specific communities at the expense of others very much depends on your perspective, if it's your home that's been destroyed or your neighborhood that's making you sick. In Stafford Beer's resonant phrasing, the purpose of a system is what it does. Infrastructural networks could be fairly described as vast constructions whose purpose is to centralize resources and agency to a small fraction of extremely privileged humans and to displace the harms to many others.

Infrastructure and Borders

In 2013, astronaut Chris Hadfield took a photo of Berlin, at night, from the International Space Station. Nearly a quarter of a century after the fall of the Berlin Wall, East and West Germany could still be distinguished. In formerly communist East Berlin, the streetlights glowed the pale orange of sodium-vapor lamps, while the rest of the city was lit with whiter fluorescents. Borders, especially national borders, are the political determinants of where our physical bodies may be located. To the extent that infrastructural systems embody sustained relationships, borders set the terms of who gets to be in them.

A century and a half or more ago, most people would live and die relatively close to where they were born. But for the small proportion who migrated, borders were effectively open and immigration was unrestricted, with some significant exceptions, like the Chinese Exclusion Act of 1882 in the U.S. It wasn't until after World War I that passports—and with them, nearly immediately, restrictions on where in the world one might live—came into common use. A key implies the presence of a lock. As a child of immigrants, I've had a passport for as long as I can remember, certainly since before I was old enough to have my own key to our family house. And as an immigrant now, I always know exactly where the keys to my home are—if they're not on my person, the metal ones are hanging on a hook by the front door, and the ones that are paper booklets are in the top drawer of my filing cabinet.

Joseph Carens, a political science professor at the University of Toronto, argues that this social order—of relatively closed borders, where citizenship is an inherited privilege—has much in common with the feudalism of the Middle Ages. Being born a citizen of a wealthy, industrially developed country like Canada is analogous to being a child of the nobility: regardless of your exact rank or wealth, you are likely to have a life of greater prospects and agency than if you were a peasant. And like feudalism, this seems like an entirely reasonable way of ordering society to most of those who are to the manor born. Carens points out, however, that there is nothing that you can say to the serf in the field that could justify why they have a different lot in life from the nobles. My parents, born and raised outside of the moat, found their way across the drawbridge so I could be born and raised within the castle walls.

The middle of the twentieth century largely saw the end of European colonialism, at least nominally, with the establishment of independent states, including India and Pakistan. Colonies were no longer subject to control by the military strength of the colonizers. (It's not quite a joke that sixty-odd countries celebrating their independence from the British every year likely makes it the most widely observed secular national holiday in the world.) But widely disparate levels of wealth and development mean that it's still possible to displace harms onto groups that can be paid to take them on, especially those that will do so most cheaply, like workers in regions with weak environmental regulations. This is especially effective if the people in the communities to which the harms are displaced don't have the option of moving from one place to another. People in areas that are more affected by the harms and negative externalities might want to relocate to the areas that have suffered fewer harms and localized more benefits. Or they may even be forced to move, as is the case for most refugees, including climate refugees. The migration itself is often challenging, even dangerous, but it's the restrictive borders that make it difficult or impossible to lawfully and safely stay when you arrive.

Like the majority of Canadians, I grew up close to the U.S. border, and while I live on the other side now, I'm still within a few hours' drive. Over the course of my lifetime, the U.S.–Canada border has hardened, in part as a response to the terrorist attacks of September 11, 2001. It's also because travel and citizenship documents have become entries in electronic databases rather than physical objects, creating a fuller record of cross-border movements and history and allowing for tighter control of passages across it.

The line between the U.S. and Canada is simultaneously invisible and directly observable; on the ground, there is a six-meter (twenty-foot) "no touching" zone, kept cleared of trees and other cover for its entire four-thousand-mile length. From the air, differences in land use are visible, since the northernmost part of the continental U.S. is adjacent to the southern-most parts of Canada. But at the same time, the border is permeable in important ways. It's nearly imperceptible in satellite photographs taken at night, for example—the lights of southern Canada merge smoothly with those of the northern U.S., with Montreal and Toronto blending into the Eastern Seaboard, or Vancouver and Seattle in the west. Electrical grids have interconnections that cross the forty-ninth parallel, and the freshwater Great Lakes straddle the border and are managed jointly by both countries. Roads and train tracks merge smoothly at the border into one network. We focus on the borders between individual countries, but the most important border lies between wealthy states with their robust infrastructure and energy systems, and ones with less collective access to money and energy. The border between the United States and Canada, like those within Europe, is just a national border. But the southern U.S. border is more than a line between two countries. It's part of an invisible ring fence around the wealthier nations of the world.

The most important example of displacing harms to communities other than the beneficiaries is perhaps the one that appears the most ephemeral: carbon dioxide and other greenhouse gases that are produced as a by-product of burning hydrocarbons. These emissions disproportionately come from countries with larger energy footprints and commensurate fossil fuel consumption. But we all share our one planetary atmosphere, which means the burden of managing the impact of greenhouse gas emissions will fall on everyone, in some way, including the residents of countries who received few, if any, benefits from that energy usage. With less access to wealth and the agency that it can provide—including infrastructural systems—those communities will likely be more affected by climate change while also having fewer resources to draw on. Even when it's a matter of survival, borders constrain the ability of individuals to move to places that are less affected or better equipped to deal with famines, drought, heat, more powerful storms, or other climate effects. Carbon dioxide in the atmosphere is allowed to go everywhere, but people are not. And as the impacts of climate change increasingly take hold (to say nothing of other phenomena, like pandemics), one response will likely be increased political pressure to "secure the borders," to make the fencing even higher and more deadly, a continuation of what's already happening at the southern border of the United States or in the responses to migrant ships crossing the Mediterranean, where nationality becomes a life sentence.

Mea Maxima Culpa

For most of my life, I thought of infrastructural systems just as these incredible technological and collective social achievements. It took me a shockingly, embarrassingly long time to recognize the inequity that is baked into nearly every one of these facilities and systems. Why on earth did it take me so long?

Part of the answer is that these infrastructural systems hide themselves from our awareness by their steady presence, their taken-for-grantedness, which means these systems of power do as well. Infrastructure isn't literally invisible—as I noted at the start of this book, some aspect of these systems is in our field of view in all but the most natural or controlled environments— but it has become so inextricable from daily life that it is only rarely *perceived*, nor does it admit any other way of living. It's just the way that the world is constituted. How could you live without running water and toilets? Or electricity? And why would you want to? I spent all of my formative years in urban Canada, in a city that was long known, a little mockingly, as Toronto the Good, and a country that aspires to "peace, order, and good government." My infrastructural birthright was extensive, and nearly all of it was public, a collective investment for collective benefit. Municipal water and sewage, electricity from Niagara Falls and my local nuclear power plants, and Bell Canada for telecommunications. It included transportation at all scales: the Toronto Transit Commission for city buses, subways, and trams; GO Transit for regional buses and trains; VIA Rail and Canadian National for passenger and freight rail; even Air Canada for civil aviation; and extensive and well-maintained highways. It's barely an exaggeration to say that these systems were built *for me*. They were largely planned and brought online or greatly expanded in the 1960s and 1970s, as the technological foundation for post-war economic growth. The human foundation for that expansion was a policy commitment to immigration, including the non-European immigration that my parents were part of. All of these infrastructural systems worked more or less in the way that they were designed to and enabled what they were intended to enable, from an economic and policy point of view. It's not that I didn't see the social norms embedded in the systems and so I presumed them to be benign. It's that since my only experience of these systems was benign, it made it hard to see them as social norms at all.

Infrastructural systems don't just hide themselves in plain sight. They can hide what came before them, like the way that highway interchanges can erase thriving communities with barely a trace. And the act of displacement is also one of hiding from awareness. I might never see the communities where the coal for power plants was mined, much less the Indigenous communities that were displaced by, well, everything—roads, railways, reservoirs, settlers, farms, subdivisions. The more expansive the networks, the farther away the harms could be. Large-scale electrical grids, for example, allow for stability, reliability, and coordinated oversight. But they also make it possible to move generating stations, and therefore the pollution they produce, far away from the consumers.

Because I was in the metropole—the in-group to which the benefits flowed—what I saw around me for much of my life was steady progress. My world was becoming safer, cleaner, and brighter, including a fair bit of harm mitigation, rather than displacement. The generating station near where I lived as a child, for all the risks inherent in nuclear power, has mostly been a steady presence on the lakeshore. Growing up, I saw the threat of acid rain recognized and addressed by the widespread use of scrubbers to remove sulfur compounds from burning coal, the rise of renewable energy, and finally the complete phase-out of coal-fired electricity plants from my home province. The hole in the ozone layer was observed, the cause was identified, and the signatories of the Montreal Protocol of 1987 agreed to eliminate the use of CFC compounds in refrigeration. It's now healing steadily, the hole getting smaller every year. Stronger environmental regulations around sewage treatment meant that Toronto beaches and the Boston Harbor have become noticeably less polluted. And I have access to entire infrastructural systems, in the form of wireless communications and the Internet, that didn't exist when I was growing up. It was perhaps inevitable that I'd only see the positive side of these systems. They're efficient! They're for everyone! These systems bring us the resources we need to survive and thrive together! That's what they're all about, right?

Spending time in India as a child certainly helped me see and be aware of infrastructure, since living in different places is a good way to realize that what's familiar is by no means universal. But it wasn't until I started actively learning more about these systems that I began to wrap my head around the implications of building out massive, worldwide networks to bring resources to specific communities. When I began to look closely at the infrastructural networks that I interact with, what I saw, over and over again, was that almost every one of them reified social injustices in some way. I slowly began to understand what would have been obvious to anyone who's ever been a member of any group that wasn't as socially central as I was, whether by income or geography or race: that you could look at these exact same systems, and understand how they function, and conclude that they exist not just to concentrate resources in a particular group in a particular place, but to do so at the expense of other people, in other places. In this light, then, all the ways of limiting where people can live, whether it's redlining or national borders or even just neighborhoods that are only navigable if you have a car, are about controlling who gets to benefit.

The Politics of Infrastructure

The collective and networked nature of infrastructural utilities can obscure power imbalances because there is often a clear benefit to scaling them to as many users as possible. Not only are water and sewage treatment plants cheaper for individual users if everyone is using the same system, but they also provide everyone with increased protection from waterborne disease. Electrical systems similarly benefit from the economic efficiencies of centralization and universal provision within a region. And transportation and communications networks become exponentially more valuable with the number of nodes. A nineteenth-century Boston landowner, even one who didn't care about the lot in life of poor people, knew that his family would be less likely to fall ill if everyone in the city had access to clean water. A twentieth-century Chicago merchant benefited from more people being connected to railway networks, to sell and ship goods to. Twenty-first-century Silicon Valley startups stand to benefit from more people with faster Internet access.

But even networked infrastructural systems are still only going to be as democratic and equitable as the societies that build and operate them. We can deduce something about the power imbalances in a society by looking at what was built and considering who benefits and who bears the harms. In more equitable communities, more effort is put into balancing, distributing, or mitigating the harms, in lieu of simply displacing them. We can also interpret it the other way around: the less effort that's made to address harms, the bigger the expression of power imbalance.

In the U.S., urban highways are a well-known example of this imbalance. The Federal-Aid Highway Act of 1956 resulted in the establishment of a federal highway trust which would cover 90 percent of the cost of new construction. Its initial authorization of $25 billion made it the biggest public works project until that time. The postwar explosion of the suburbs, the growth in car ownership, and this federal funding kicked off a country-wide highway building boom. Across the U.S., interstate highways were intended to facilitate trade, hence the federal money. But within cities, the main goal of highway planning was to ease commuting by private automobile from the rapidly growing suburbs into the central business district. This might sound uncontroversial, but in the 1950s and 1960s, when the new highways were being sited, planned, and constructed, women were mostly excluded from the workforce and Black residents were largely unable to purchase homes in suburbs because of the practice of redlining, in which banks would refuse to give mortgages to Black households except in a small number of "acceptable" neighborhoods. The end result was that highways in American cities were planned and designed by White men, largely for the benefit of White men, the group that benefited most from being able to drive between the suburbs and downtown. To this day, many U.S. cities still have a hub-and-spoke commuter model in local transportation networks, including public transit. Here in Boston, rather than cardinal directions, subways and commuter trains head "inbound" or "outbound"—and, as in many cities, it's notably easier to go down a "spoke" to the hub than it is to travel between spokes. The commute-centric model has long been at odds with research that shows that caregivers, still mostly women, are more likely to chain together segments than to make a regular in-and-out trip or, more generally, that residents need to be able to get around their own neighborhoods. Half a century after these highways were built, with the advent of the Internet and a pandemic that normalized working from home, the whole idea of a daily commute from the suburbs into the downtown core is being reconsidered.

Building out new highways meant demolishing and dividing communities, and bringing noise and air pollution to nearby residents. In the U.S., this pattern of urban highways being built through African American neighborhoods, as well as through neighborhoods that had large immigrant populations, is nationwide. It was considered sensible and justifiable by the planners and engineers because those were usually the neighborhoods with the lowest property values, requiring the least amount of compensation to be paid to expropriate the land for public use. In many cities, these neighborhoods were even considered an urban "blight" of unsuitable or undesirable housing stock, which meant that planners considered it better for the residents to live elsewhere, regardless of what the residents themselves thought of their own homes or where they could move. In both Canada and the U.S., similar arguments were made for displacing Native American communities in order to construct dams and reservoirs—it was "low-quality" land, so it required only minimal compensation to be taken by eminent domain, and the inhabitants should be grateful for the opportunity to relocate to "better" places.

The Southwest Corridor in Boston, with the noisy trains, was intended to be the Southwest Expressway but its construction was canceled in 1970, largely as a result of resistance in the local community. They worked with local planners who had observed what had happened with other new highways in Boston. Interstate 93 downtown was built through Chinatown, and Interstate 90, just east of the city proper, displaced a predominantly Italian neighborhood. By then, there had been other successful freeway revolts all across the U.S. and Canada. The cancellation of the expressway in Boston was a stepping stone to getting the entire Federal-Aid Highway Act amended so that it could be used to pay for transportation modes besides just highways, including the Orange Line subway that was eventually built in the Southwest Corridor.

Infrastructural systems make it possible to accrue resources and agency. Simultaneously, and not coincidentally, over the course of the twentieth and twenty-first centuries politically marginalized groups have fought for and gained power, and used their increased civic agency to advocate for equitable access to the benefits of public infrastructure and to resist being forced to bear the brunt of harms. The fight over the Southwest Corridor in Boston was an outgrowth of the U.S. civil rights movement. Globally, we see it in the recognition of women's rights and the political independence of former colonies. Democracies moved toward pluralistic, multicultural, multiracial societies (although not monotonically, nor without resistance). The globalization of infrastructure in the twentieth century—containerization, aviation, and telecommunications—means there is no longer an invisible other onto whom the harms can be displaced: the voices of affected communities can be and are heard. The Janus-faced nature of infrastructural systems—of both centralizing benefits and displacing harms—is increasingly visible, understood and, rightly, challenged.

Where Does This Leave Us?

The power, utility, and value of infrastructural systems are largely due to their collective nature. They are the most effective way that humans have ever devised to empower people in their daily lives. It's why systems like water, sewage, electricity, and communications are developmental priorities in communities that don't have them, wherever in the world they might be, and why infrastructural systems often follow trajectories that lead to their becoming political rights. They span the globe and are marvels of cooperation and technology, allowing a large fraction of the world's population to live lives that would be unthinkable to our ancestors, as recently as our great-grandparents.

That also means that those of us with good, consistent access to these basic utilities are completely dependent on these systems to live our lives which makes us extraordinarily vulnerable to their failure. We should be seeing them, celebrating them, and protecting them. Instead, these systems have been invisible and taken for granted.

This is in part because infrastructure hides itself. "Infra" literally means "below"—these systems are not just metaphorically but often literally invisible, in our walls and floors, under our roads, in a restricted area, or surrounded by a level of inattention that the science fiction author Douglas Adams satirically described as a "Somebody Else's Problem field." They're cognitively hidden from us, because well-functioning systems are just always there, to be taken for granted, calling for very little of our conscious attention. And the harms associated with these systems are even more hidden, displaced by means of networks and energy. The ever-increasing power of infrastructural systems to facilitate that displacement means that there are few incentives to even attempt to mitigate harms, except the most egregious, when it's long been an option simply to find another place or community that can take on those harms, either by payment or through force (or both). Seeing and understanding infrastructural systems is not just about recognizing their value. It also means squarely facing these entrenched inequalities.

The prison abolition scholar Ruth Wilson Gilmore uses the term "organized abandonment" to describe the systematic removal (or lack of provision) of state-provided services from marginalized racial groups. The consequence of that lack of support is reduced civic agency: the inability to fully participate in civil society, including being denied educational and economic opportunity. I think of well-maintained infrastructural systems, particularly public ones, as the opposite of Gilmore's organized abandonment—universal provision of systematized, physically instantiated state support. Inequalities in provision matter, and a failure to provide these services, or the failure of these services, is a form of organized abandonment. Without infrastructural systems, we can't fully function as individuals or as communities. And we are *all* vulnerable to their disruption.

I've been living in the U.S. for two decades now, and I've observed the organized abandonment in the wake of major hurricanes, particularly New Orleans after Katrina in 2005 and again in 2017 in Puerto Rico after Maria. Even after Hurricane Sandy in New York, there were differential responses between downtown Manhattan and less wealthy neighborhoods. Maybe one of the saddest observations that I've made as an immigrant to the U.S. is that what happens to the most marginalized groups in the country has often been a harbinger of what less marginalized, more socially central groups would soon be facing. I thought about this in February 2021 when a winter storm resulted in a multiday blackout across Texas, leading to days of freezing misery and hundreds of deaths.

Whether gradual or catastrophic, localized or widespread, the failure of infrastructural systems is a particular type of organized abandonment. Any significant loss of these systems from our lives is recognizably dystopian and, to borrow from the science fiction author William Gibson, that future is already here, it's just not very evenly distributed. A failure of infrastructural provision might be caused by something as mundane as undermaintenance, as out-of-the-blue as a solar storm, as infuriating as poor governmental oversight, and as all-encompassing as climate change. And none of these threats are specific to the U.S. or any other single country. If we want to protect our infrastructural systems, we need to recognize and begin to come to terms with these threats.

Seven

HOW INFRASTRUCTURE
FAILS

Threats to infrastructural installations are a staple of thrillers, whether in blockbuster movies or cozy mystery novels—even one of Louise Penny's Inspector Gamache novels, mostly set among the quirky residents of the tiny, fictional village of Three Pines, outside Montreal, involved a subplot in which attackers plan to drive trucks loaded with fertilizer bombs to La Grande, the massive hydroelectric dam complex in northern Quebec. Unsurprisingly, infrastructural facilities have long been targets in warfare, but that became especially true in the twentieth century as these systems grew in importance. During World War II, the British sent a specialized air squadron, later known as the Dam Busters, to carry out a coordinated attack on the industrially important Ruhr valley of Germany, breaching the dams and destroying hydroelectric power stations. A decade or so later, late in the Korean War, similar attacks by American bombers in North Korea destroyed 90 percent of its power generating capacity and caused a countrywide blackout that lasted two weeks.

These kinds of attacks—intentionally depriving civilians of survival necessities, including food and electricity, as well as targeting dams, nuclear power plants, and other systems that can release dangerous amounts of energy if damaged—are now violations of the Geneva Convention and so considered to be war crimes. Infrastructural systems continue to be targets of physical attacks, of course, because those same characteristics make them high-value targets, especially for terrorists and other nonstate actors. A 2013 attack just outside San Jose, California, in which multiple coordinated attackers fired high-powered rounds into the transformer banks of the Metcalfe Transmission Substation, focused attention on the risk to the U.S. electrical grid. Those attackers were never identified and so their motivation remains unknown, but in recent years, particularly since 2020, the electrical grid has been specifically highlighted as a potential target by far-right extremists with anti-government, white supremacist ideologies, and there is some evidence that these attacks have been increasing in frequency. In general, these sorts of attacks result in relatively little disruption because power can be quickly routed around the damage. But that's not always possible, and a December 2022 attack led to a multiday blackout for tens of thousands of people in North Carolina. What keeps these systems safe isn't that they're rare, hidden, or protected—it's that they're either just taken for granted or considered benign. But if extremist groups decide that infrastructural facilities are illegitimate and target them, they constitute what computer security researchers call a large "attack surface," with many points where they can be threatened.

Deliberately and maliciously disrupting the electrical grid on a medium to large scale is surprisingly challenging. Paradoxically, the same thing that makes the electrical grid and other infrastructural systems vulnerable—that they are distributed across and embedded into the landscape—also means that they are generally designed for oversight and redundancy. This can limit the effect of bad actors, and failures cause downstream and localized effects. Widespread disruptions are most often caused by widely distributed events, as when severe weather downs power lines over a large region. Somewhere around 15 percent of all power outages in the United States are caused by notably small but widely distributed actors, although it would be unfair to call them malicious, since they're mostly squirrels, as well as raccoons, birds, and other animals who picked the wrong wire to perch or chew on. It's been estimated that bringing down the whole U.S. electrical grid (which includes nearly ten thousand generation sites and two hundred thousand miles of power lines) would require simultaneously disabling nine of the thirty or so largest high-voltage transformer stations that are the critical links between different regions across the country. Each station is about the size of a small house, but they're like the Death Star—vulnerable to targeted damage from someone who has a basic idea of how they function. Nevertheless, taking out all of them would require a large-scale, coordinated attack.

Especially in the last two decades, the idea that critical infrastructure needs to be protected, by force if necessary, has become well established. One bright midsummer morning a few years ago, I returned to the nuclear plant in Pickering where I spent so much time as a child. I first went to the visitor center, and then for what turned out to be quite a short walk around the outside of the installation, because I was quickly intercepted by two conspicuously armed guards, presumably after being spotted on surveillance cameras. They asked me what I was doing, insisted on taking my phone and deleting the few photos I had taken from outside the fence, and then politely but unceremoniously put me into their car, drove me back to the parking lot, and stood watch to make sure I got into my own car and left. It was a bit of a novelty for me—as a nice, middle-aged lady and an overly enthusiastic engineering professor, I'm not usually treated as a security threat—but it certainly also gave me some uncomfortable insight into the degree to which the facility was expecting and was prepared to respond to physical intruders.

Virtual intruders and other network-based cyberattacks present a much more amorphous and evolving risk to infrastructural systems. Colonial Pipeline operates one of the major pipelines for gasoline, diesel, and jet fuel between Texas and the southeastern U.S., and in May 2021, its billing systems were targeted by the hacker group DarkSide in a ransomware attack.

Even though the hack nominally only affected administrative functions, Colonial shut down fuel provision entirely because they couldn't be sure the operational software for the pipeline wasn't also compromised. (It's probably not incidental that, without a working billing system, they had no way of charging customers.) The result was severe supply disruptions and fuel shortages for a week, affecting gas stations, trucking, and flights across the region.

Operational control of infrastructural systems is always the primary concern. My students and I visited our local water treatment plant in Needham, Massachusetts. The tiny facility has only a handful of staffers, and they showed us how they can monitor the water treatment process remotely, from wherever they are, and can address issues without being physically at the plant. While it makes sense to connect infrastructure facilities to the Internet to help operators maintain continuous oversight and respond to emergencies, anything that can be controlled from the larger world can, almost definitionally, also be compromised remotely. I thought about the Needham plant when I heard about an intrusion into the control software for water treatment in the town of Oldsmar, Florida (population 15,000), in February 2021. The plant operator on duty noticed a cursor moving around the screen, and then saw a number on the display jump suddenly from 100 to 11,100. It was the desired concentration in parts per million of sodium hydroxide, used to balance the pH of water as part of the treatment process. Commonly known as lye, sodium hydroxide is a poisonous alkali that can burn skin, and someone had just tried to make the concentration in Oldsmar drinking water a hundred times higher. The alert operator immediately locked the intruder out and reverted the system to the correct concentration, although fail-safes would've prevented the doctored water supply from leaving the plant in any case. As with the Metcalfe attack, the intruder was never identified and their motivation is unknown, but tampering with a public water supply in any way, for any reason, and to any degree would never be considered unimportant or benign. The attack was widely seen as a wake-up call to address the cybersecurity of water utilities, including the question of whether they should be connected to the Internet at all.

Physical threats to critical infrastructure, especially charismatic mega-structures, get all the screen time; cyberthreats, with their inherent uncertainty and rapid speed of change, keep security researchers up at night. It's headline news when infrastructural systems fail. Their ubiquity makes them vulnerable because it's challenging if not outright impossible to protect them as a whole. A reservoir for drinking water, for example, is usually also part of a local watershed, so it can't be physically isolated, much less relocated, and it needs to be protected from mundane threats like accidental spills as well as from deliberate attempts to taint it. But the flip side of that ubiquity and distribution is that the most significant threats to these systems as a whole aren't coming from individual bad actors, terrorists, disgruntled former employees, or script kiddies out for lulz and bragging rights. I don't want to downplay the effects that an attack on a single facility, like a drinking water source, might have. But the most dangerous threats—the ones that keep me up at night—are those that function on the same scale as the systems. Physical and military threats—active human intent to cause harm— are only some of the possibilities. There are also large-scale threats to infrastructure that are *unintentional* effects of human actions, and ones that are the result of human *inaction*. And there is also a class of threats that aren't a result of human intent or action at all.

Gray Swans and Blackouts

Fictional threats aside, the dam at La Grande in northern Quebec has never been breached, but one cold March morning in 1989, a cascading failure disabled the hydroelectric power station of the massive system in less than a minute, taking half of the province's power supply down with it. The resultant nine-hour blackout included the city of Montreal and had ripple effects extending into New York and New England. It was the most recent and severe return of a phenomenon that had been observed more than a century earlier, but this time it had very different effects.

An hour or two after midnight on September 2, 1859, gold miners in the Rocky Mountains were woken by a sky so bright that they started eating breakfast and getting on with their day. They were witnessing what became known as the Carrington Event, an enormous geomagnetic storm caused by the interaction of a massive solar flare with the earth's atmosphere, which produced striking auroras visible as far south as the Caribbean. But the powerful magnetic fields of a solar storm can also induce electrical current in wires; during the Carrington Event, telegraph operators reported receiving shocks from their equipment, and subsequent strong storms have even started fires.

A storm like this, in the middle of the nineteenth century, is a historical curiosity. Now, in the twenty-first, it could be a catastrophe. Power lines crisscross the continents, and the planet is enmeshed in a web of telecommunications networks and surrounded by a fleet of satellites that transportation systems, especially aviation and shipping, depend on for navigation. All these infrastructural systems would be vulnerable to the fluctuating and powerful magnetic fields of another Carrington Event. A strong storm like this caused that 1989 blackout in Quebec: the induced currents tripped a breaker in a substation, leading to a rapid cascade of failures and the loss of control over nearly ten thousand megawatts of electrical power.

Similar but milder events are frequent and are part of what's known as "space weather," but a severe solar storm would be a "gray swan" event. Nassim Nicholas Taleb popularized the terms "black swan" for a high-impact event that can't be predicted ahead of time and "gray swan" for an event that's extremely rare and whose exact timing is unpredictable, but which can nevertheless be expected to occur and to have a tremendous impact when it does, like a global pandemic. The next Carrington Event would have the potential to knock out the grid for millions of people, disable satellites, and cripple telecommunications, with a financial cost that would be likely to reach into the trillions of dollars for the U.S. alone.

Quebec got a heads-up in the 1989 blackout, as its system is particularly sensitive to solar storms due to the geography of the province. It's closer to magnetic north, and long power lines running over the nearly exposed granite of the Canadian Shield (rather than over deeper soil) are especially vulnerable to the kinds of induced currents that destabilize the electrical grid. But Hydro-Quebec is also unusually well positioned to respond to and prepare for this threat, and has already begun work on hardening its facilities and building out systems to minimize the disruption caused by a severe storm. It generates a tidy profit (more than 3.5 billion Canadian dollars in 2021 and in 2022, some of which came via my own utility bill, in Massachusetts) and, as a publicly owned power utility, it has a mandate to serve the residents of the province into the indefinite future. That means that Hydro-Quebec has both the resources and the responsibility to protect against gray swans—it's enough to know that one will arrive even without knowing exactly when.

The U.S. response is predictably slower and more piecemeal, given that its infrastructural systems are a patchwork of public and private providers. Still, the consequences of the next Carrington Event are well understood and there is federal funding and industry-specific guidance to make facilities and systems less susceptible to disruption when it does arrive. The Space Weather Prediction Center, part of the National Weather Service, monitors solar flares, and makes predictions, providing daily forecasts to the affected industries.

Will Durant is credited with the adage that "civilization exists by geological consent, subject to change without notice." Major earthquakes are gray swan events, and so are volcanic eruptions. Mount Rainier, whose majestic cone is easily visible from Seattle, is an active volcano. Any such event is likely to have a huge impact on the region. Emergency preparedness is in large part about planning for survival in the face of significantly damaged or otherwise inaccessible or unusable infrastructural provision. The impacts of natural disasters, like a major earthquake or a volcanic eruption, depend heavily on the way that basic services have been designed, built, and maintained. More than anything else, poorly maintained infrastructure is much more likely to fail, and to be much more difficult to repair, under the stress of a major disaster.

That brings us to another way infrastructural systems fail: via chronic, slow erosion, the kind that might be almost imperceptible—until it's suddenly not. These failures are less a result of human action than human inaction.

Under Our Feet

On their first day of college, I take my engineering students for a walk into the past.

The college is currently only a few years older than our undergraduates— the campus was mostly built in 2002 and is very modern with pale brick, high-ceilinged atria and sheer glass walls. Our class walks out of the main academic building, down the driveway that leads to the main road, and onto an unassuming grassy berm alongside a patch of wetlands. As we continue, it becomes increasingly clear that we are walking on *something*—a low, linear mound, far too regular to be natural. We follow it to the main road and jaywalk across (sorry, parents) as the suburban drivers politely pause for the small crowd of students. On the other side, there's a well-defined trail, and we keep going as it crosses the access road to the municipal solid waste facility. By now it's evident that we are walking at the apex of an arch, and a few minutes later we arrive at a small brick and stone building, set athwart the path with its door padlocked shut and looking distinctly unmodern. My curious students circle it, noting the stream crossing below, and if they haven't already, this is the moment when they clock that we've been walking on top of an underground tunnel. It's the Sudbury Aqueduct, built during the 1870s to carry water from the western suburb of Framingham into Boston's Chestnut Hill reservoir, and is part of the Massachusetts Water Resources Authority (MWRA), which supplies the region with much of its drinking water as well as managing wastewater, including the Deer Island plant that I visited.

I take my students out to the aqueduct for a few reasons. The first is to help them learn to *see* the infrastructure in the world around them—that cutting across our contemporary campus is a century-old water tunnel, but that it's only apparent if you know what you're looking at. At the end of class, I point them toward the informational display about the Sudbury Aqueduct that's tucked into a corner of campus behind their dorm and I send them on a treasure hunt to find the nondescript blue-painted pin that marks where the buried channel for potable water crosses under the road, sidewalk, and playing fields. I receive a lot of selfies from my students, pointing at the pin and usually looking delighted. They've only been on campus for a week or so and have already found something that many of their classmates won't notice in four years of college.

Learning about the Sudbury Aqueduct also helps them begin to get a sense of the longevity of infrastructure, the way that older structures form an underlayer for even the newest facilities. This particular aqueduct isn't even in regular service. It's kept in good repair so it's available in emergencies—just a few years ago, it was brought into service when a pipe broke in the next town over. I also want my students to begin to realize that this aqueduct, with its charming signs and careful maintenance, is something of an anomaly, at least in the U.S. Unlike my enthusiastic and high-achieving first-year engineering students, the infrastructural systems around them mostly get failing grades.

Every four years, the American Society of Civil Engineers (ASCE), the governing body for the profession, does a complete review of the state of U.S. infrastructure. It covers all fifty states, and they consider transportation systems, including aviation, bridges, roads, transit, rail, ports, and inland waterways; dams and levees; drinking water, wastewater, solid waste, and hazardous waste; the energy sector; schools; and public parks. Each system receives a score, and then the ASCE distills all their findings down to a single overall letter grade. In 2021, the most recent report as of this writing, they gave U.S. infrastructure a C minus, up slightly from the D plus in 2017. It's a grade that would be unimpressive for a student, but it's downright dismal for these systems on which we depend for our lives and livelihoods. In their scheme, a C means that "some elements exhibit significant deficiencies in conditions and functionality, with increasing vulnerability to risk," while a D is used for systems in "in poor to fair condition and mostly below standard, with many elements approaching the end of their service life. A large portion of the system exhibits significant deterioration. Condition and capacity are of serious concern with strong risk of failure."

Vulnerability to risk. Strong risk of failure. This isn't due to hostile actions or a cataclysmic natural event. It's largely due to neglect—lack of investment, care, and maintenance, and all that follows from that. If a solar storm is a gray swan event, then chronic lack of maintenance is like "red termites"—a risk from a threat that is continuous but mostly invisible or ignored, in the same way that insects chewing on wooden supports can go unnoticed until the floor collapses.

Things Fall Apart

Mid-morning on the Saturday of a holiday weekend in August 2018, in the pouring rain, the cables along one side of a span of the Morandi Bridge in Genoa, Italy, snapped and set off a tragic sequence. No longer held suspended, that side of the roadbed began to tilt, putting even more load on the remaining stays on the other side, which gave out in turn. The roadbed, carrying dozens of cars and several trucks, dropped fifty meters into the river and the industrial area below the bridge. Moments later, the central tower collapsed. Forty-three people were killed.

The bridge was designed and built in the 1960s, and was considered at the time to be the masterwork of its namesake, the Italian civil engineer Riccardo Morandi. He pioneered a technique of encasing the support cables in concrete, which he believed would serve as an inert barrier and protect the steel from the elements. The Pantheon in Rome, after all, was built two thousand years ago and is still standing, the largest unreinforced concrete dome in the world. But just a decade after the bridge opened to traffic, Morandi realized his mistake: the concrete was already deteriorating and, what's worse, the embedded cables were beginning to show signs of corrosion. Bridge inspections over the next three decades revealed increasing amounts of damage and additional cables were installed on several pillars (but not, crucially, the ones that collapsed). In the spring of 2018, just after the bridge's fiftieth birthday, the private company that administered it put out a call for tender for a €20 million structural upgrade. A few months after that, the Morandi Bridge collapsed.

I've broken an arm three—three!—times. Once when I wiped out on my bike, once when I fell over while ice skating, and once while I was on roller skates, at an introductory roller derby session (needless to say, that was the end of my derby-girl dreams). My penchant for breaking bones is particularly funny, to me at least, because I spent most of my twenties in graduate school researching how the mechanical properties of bone tissue change with age, stress, and disease—in fact, when I fell on the ice, I went straight from the rink to the hospital that housed the lab where I worked, and was seen by an orthopedic surgeon I was collaborating with. My graduate work was mostly in materials science and engineering, where you learn that very few things fail catastrophically because they are suddenly overloaded past what they are designed for, the way my radius and ulna were. The kinds of gradual failures that I was studying are much more common, both in the body and in engineering structures. Structures frequently fail in fatigue, where low levels of stress lead to creeping damage that accumulates over time. Without careful monitoring, that damage goes unobserved and therefore unrepaired. In bones, fatigue failures show up as stress fractures, which are common in athletes and military recruits. As we walk, run, and jump around, our bone tissue accumulates microscopic cracks, which are quietly repaired by specialized cells when we're at rest. Overtraining, like repeated long marches while carrying a heavy pack—more damage, with less recovery time—can overwhelm these cellular mechanisms. The previously unnoticed microcracks merge to become a painful stress fracture.

Another common mode of failure is the opposite of fatigue: rather than low loads resulting in damage accumulation, gradual degradation of the material means that it eventually can no longer sustain even normal design loads. In bones, this process begins in early adulthood, after our skeletons finish growing, and continues as we age. We gradually lose bone tissue over time and if enough of the material is lost, we become susceptible to osteoporosis and fractures, typically of the hip or in the spine.

In engineering structures, a gradual loss of structural integrity is often due to corrosion. The material of the structure degrades as a result of exposure to its environment, usually air and water, but also salt and sometimes even other metals. The best known example of corrosion is the friable brown rust of iron and steel, but that's just one member of a large and diverse family. Good design can minimize or control corrosion, and the exact approach depends on the use case and the cost—we use surgical steel for medical implants and stainless steel for cutlery, for example, while housing nails are plain steel that's coated or oxidized. But corrosion is rarely eliminated entirely. Left unchecked, the end result is the same as fatigue failure: the structure fails unexpectedly during normal use.

For well-designed engineering artifacts, from bridges to locomotives to soda cans, failures are most likely to result from these sorts of gradual, low-level damage that occurs under "normal" conditions and accumulates over time. Engineers know that this kind of mundane failure is usually the biggest risk. But there's a real cognitive shift here, a gut-level professional understanding that just because something looks and functions like it should, it doesn't mean that you can assume that everything is fine and will stay that way.

There is no getting around maintenance. Things fall apart. Fatigue accumulates, cracks grow, steel corrodes, and squirrels chew through wires. The universe tends toward disorder, inexorably, unless we put energy in to hold the line. The first and the third laws of thermodynamics tell us that energy can't be created or destroyed, and that all energy eventually ends up as heat, but this is the second law of thermodynamics: that entropy increases. In real, complex, material systems, the continued functioning of all engineering systems, including infrastructure, depends on routine inspection and maintenance.

The good news is that gradual failure is, well, gradual. The aluminum alloy of airplanes, for example, accumulates damage in a way that's similar to bone tissue, with cracks forming and then growing longer and longer because of low-level but ongoing stresses, like pressure differences and vibration during flight. In a well-designed system, any cracks in the structure of the plane can be identified, observed, monitored, and repaired before they become critical failure points, which is part of why there are rigorous and carefully logged maintenance schedules in aviation, and similar strategies are used for all sorts of critical systems.

But without careful attention to repair, systems degrade. It seems like this kind of gradual, accumulated damage should lead to chronic low-level failures, and it absolutely does. In the U.S., after decades of neglect, water pipes are now being replaced at the rate of 1 to 5 percent per year on average, but even this means it might take a century to replace a whole system. The pipes are designed to last decades but not centuries. Many water transmission and distribution systems in the U.S. are getting up there in age— those that came online in the middle of the twentieth century are now pushing a century of use, and some, like the Sudbury Aqueduct, were built the century before that. One consequence of these aging pipes is that around 15 percent of all clean drinking water in the U.S. is lost to leaks.

Inadequate maintenance can also lead to acute failures, where systems and structures can work fine until they suddenly don't. Often this is because there is a redundancy or a safety margin that is gradually eroded over time. Bridges, overpasses, and other civil engineering structures, for example, are designed with significant safety factors, which means it often takes a lot going wrong before they fail catastrophically (and is why those failures are still, thankfully, a rare and newsworthy event). Routine inspections and maintenance are what help to ensure that the safety margin remains intact, including by revealing issues with the design or construction. The Morandi Bridge, for example, failed because of a design issue related to corrosion—it was underdesigned for the load it carried, with limited redundancies, because of a misguided belief that the cables would be protected from the salt air and industrial pollution of Genoa. Routine maintenance led to the recognition that this was an issue, although part of why it wasn't addressed in time was because encasing the cables in concrete made it a challenge to get a clear understanding of the degree to which they were weakened. The Interstate 35W bridge, which collapsed over the Mississippi River in downtown Minneapolis during evening rush hour in August 2007, was revealed by post-failure investigation to have had a more conventional design flaw— a gusset plate connecting the beams of the bridge to its support columns was too thin for the load it carried, and it failed by ripping across a line of rivets. But it was already known that the bridge wasn't in great shape—it had been rated as "structurally deficient" in 2005 due to corrosion issues, putting it among the bottom few percent of the 100,000 or so most heavily used bridges in the U.S. Follow-up inspections the next year found problems related to cracking and fatigue, but remediation of the forty-year-old bridge was planned and shelved because of concerns that the plan for retrofitting would weaken it further, which meant that it instead had to be replaced. It's painfully ironic that a contributing factor to its collapse was maintenance work on the roadbed—four lanes were closed for resurfacing, and there were around 250 tons of construction equipment on the bridge.

Examples like these help explain why I think of undermaintenance of infrastructural systems, usually as a result of underfunding, as an actual threat: a "red termite" risk. Like termites, these issues can be widely distributed but still easy to miss or to ignore, especially in the early stages of infestation. But, as with termites, that lack of vigilance and oversight can result in not just chronic or low-level failures but also catastrophic failures of physical systems. Bringing down infrastructural systems doesn't require hackers or munitions. It just takes failing to provide the expertise, care, and investment that keeps them running continuously and safely. This was the case with the failure of the Edenville Dam in Michigan in May 2020, which led to the evacuation of ten thousand people at the height of the coronavirus pandemic and an estimated $200 million in damages. As a privately owned "cash register dam," the costs of maintenance were meant to be covered by sales of the hydroelectricity generated, but for a decade and a half, its operators refused to make the mandated repairs and upgrades to a spillway that had been found to be inadequate. While this had eventually led to the loss of its electricity generation license (and hence, income) and a community-based plan to purchase and operate the dam, the struggles over regulation and ownership meant that the repairs weren't completed before that sodden spring. Neglect a system badly enough, or long enough, or incautiously enough, and it will be at risk of failure. Any sufficiently advanced negligence is indistinguishable from malice.

This runs in the other direction too: the neglect of infrastructural systems can be weaponized. In 2018, there were reports of outbreaks of waterborne diseases, including typhoid, hepatitis A, and dysentery, in areas of Syria where the population was considered to be hostile to the rule of President Bashar al-Assad. Assad's military strategy for this region included not just conventional warfare but also arresting water treatment plant engineers and maintenance staff, as well as withholding the chlorine needed to treat water supplies—biological warfare by the expedient of making sure that the quiet, daily work of operations and maintenance didn't take place.

The Challenges of Maintaining Networks

There's a line that's become a bit of a cliché in the technology world, often used by start-ups that are in the process of defining what they're making and what their business model is: "We're building the plane while flying it." The metaphor is vivid precisely because of how impossible it is; even routine maintenance of airplanes is something that we'd all prefer to have happen on the ground, thank you very much. But because we depend on and use infrastructural utilities around the clock, all the time, there is no time when they are "off the clock." And without a clear understanding and appreciation of the value of maintenance and the work being done, even planned down-time for maintenance is a disruption and can seem like a mini-failure to users. (My friends who worry that I don't get enough rest or relaxation have taken the opportunity to point me to an industrial safety sign that reads, "If you don't schedule time for maintenance, your equipment will schedule it for you.")

Not only do we take well-functioning, long-lived systems for granted, they are often *actually* hidden from view, with a degree of accessibility that's in proportion to how frequently we expect to interact with them. In buildings, that means that light switches and outlets are conveniently positioned for use, fuse boxes or circuit breakers are out of the way but accessible, and the actual wiring is mostly embedded in the walls. At the municipal level, the "infra-" in "infrastructure" is frequently literal, with pipes, ducts, and wires buried below street level. New infrastructural networks tend to be routed along existing networks when possible. Telegraph wires were strung along train tracks and phone lines were installed alongside power. Today, water and sewage pipes, electrical lines, natural gas, and telecommunications systems are typically run alongside or under existing roads, in part because they're a clear public right-of-way. But this multiplexing of infrastructural networks has a number of consequences for maintenance. For a start, any kind of utility work is likely to disrupt other systems—if nothing else, the road above it. Especially in cities, it can even be hard to find the relevant bits, after decades of accumulation and incremental changes. These systems are dug up and exposed when they need repair, and the repaired version isn't necessarily the same, or in the same position, as the original. Locating buried infrastructure is such a clearly recognized skill that there's even an International Utility Locate Rodeo, where workers compete to demonstrate their skill at finding belowground pipes and conduits.

Shutting down services or roadways for scheduled repairs or upgrades is annoying and disruptive, but avoiding it is a bad idea, since the disruption of planned maintenance is usually rather less than that of unscheduled repairs, if only because systems rarely fail at convenient times. It's the biggest argument for getting them right in the first place and then maintaining the hell out of them, so they can be kept hidden away and then built on or extended as needed, on a planned timeline, without having to rip them up and start over.

Redundancy is a hallmark of infrastructural systems in part because it provides for maintenance and downtime, but it typically requires both an upfront and an ongoing commitment of resources. The Cambridge water treatment plant has *three* parallel lines through it, each one capable of treating the entire city's water supply—that means that even when one line is offline for planned work, there's a backup system if one of the two remaining lines fails unexpectedly, as well as an interconnect with the larger MWRA system, in case of a major disruption with either. The San Francisco ferry system that I love is relatively lightly used in comparison to either the roadways or the BART train system, but in the case of a major earthquake that could damage or close bridges and tunnels that cross the bay, a scheduled ferry service ensures that there will still be a nonterrestrial route from the city proper to the East Bay region.

Network redundancy is valuable in more mundane emergencies too. Multiple transit routes in cities, for example, provide flexibility to commuters. When I was living in London, I often took the Overground for my commute into the city center from my flat in the northern part of the city. I liked taking the train because it was fast and comfortable, but it was also relatively infrequent. One morning I arrived at the station to realize that I'd forgotten my transit card. On my short walk home to get it, I pulled out my phone, launched a transit app, and figured out an alternate routing, via buses and the Tube, arriving at my destination only slightly later than I had planned. Particularly as someone who's spent a lot of time in U.S. cities where so often there are few good alternatives to driving, this kind of deep transit redundancy felt like the transportation equivalent of having three lines through the water treatment plant.

Sustaining Infrastructure Is Sustaining Relationships

Deep in the Andean mountains, about a hundred miles from Cuzco, the Q'eswachaka Bridge is suspended high above the Apurímac River, spanning a gap of 38 meters (about 120 feet) between stone abutments. Once part of a network of routes that knitted the Incan empire together through the rugged landscape, the footbridge still serves as an important local connector. It's been in use for nearly seven centuries, despite being constructed of a humble material: grass. Every year, four neighboring Quechua communities come together at the bridge for a three-day festival, bringing with them the materials and expertise needed to rebuild it. Long grasses are woven into bundles, then cables, and then braids of cable. They cut the previous years' bridge free from its moorings, letting it fall into the river below. Under the supervision of a bridge master, teams weave the new bridge into place, starting at opposite sides of the canyon: first the cables, then the deck, and finally the side panels that serve as railings.

There's a philosophical puzzle known as the Ship of Theseus. Following his return to Athens after slaying the Minotaur, Theseus' ship was preserved as a memorial, reportedly for centuries. But once a year, it was sailed to Delos, on a pilgrimage to honor Apollo. The conundrum is this: if the ship is repaired over and over again to keep it seaworthy, to the point where every individual element of the vessel, every timber and line and sail, has been replaced by a new equivalent, is it still the same ship? Infrastructural systems can be real-life examples of the Ship of Theseus. We don't have much difficulty in saying it's the "same thing," whether it's a suspension bridge that is replaced in its entirety every year or a network of underground pipes that's been gradually but completely replaced over decades. One way of thinking about this is that infrastructural systems are defined not by what they are or what they are made from, but rather by *relationships*: between the various parts, and how they interact with other elements of the network, and how we interact with them. When the routers that provide wireless Internet service at a workplace are replaced with updated ones, many users wouldn't even notice—from their perspective, all that matters is the relationship between computing devices, wherever they are in the building, and the outside world, and the way they can continue to take it for granted.

On a drizzly gray London morning in 2017, I put on a hi-viz tabard—a fluorescent green vest—and a hardhat, and went through the gates of London Bridge Station, on the south bank of the Thames, while it was in the middle of a massive renovation project. It stayed open to passengers throughout, which meant that the construction work had to be scheduled around the trains and the trains had to be routed around the construction. Two at a time, tracks were taken out of service, everything around them was removed and replaced, the rails were rebuilt, and then the crews moved on to the next pair. What it really highlighted for me is that with infrastructural systems, there isn't really a bright shining line between maintaining existing systems and building them anew. While bridges might be carefully maintained and inspected up to the point where they are replaced, in many cases, maintenance is neither just repairing what's already there nor is it wholesale replacement. Instead, it's renewal, replacement in an incremental, planned way, much like the way that accumulated damage in bone is repaired by the body with new tissue. The London Bridge Station, a newly rebuilt Q'eswachaka Bridge, or even something like the new Zakim Bridge, all have largely the same set of relationships to both the people who use them and the area around them, whether they're train passengers south of London, foot traffic in the Andes, or drivers coming in and out of Boston on Interstate 93.

Infrastructural systems require a sustained investment of resources. That investment can be built into the upfront cost (like building a water treatment plant in triplicate), into operational and maintenance costs, and into scheduled disruptions. And it's built into renewal and replacement. In most cases, the cost of having systems that we depend on and *not* paying for them to be robust and well maintained is even higher, in part because it's measured not just in money but in the societal costs associated with their failures. The longevity of infrastructure and the sustained relationships that depend on its presence mandate that there is a smooth segue from planning and design, through operations, to maintenance during use, to renewal, to rebuilding, to a total replacement.

But there is one final piece to this, which is that these systems can't be static. Even if they're thoughtfully designed, carefully built, and conscientiously maintained, that wouldn't be enough, because the world that they are embedded in, and our understanding of it, changes with time.

Responding to a Changing World

Like Lewis Carroll's Red Queen in *Through the Looking-Glass*, maintenance and repair of our infrastructural systems can be about running continuously to stay in place, perpetually fighting the good fight against the forces of entropy. But we don't always want to stay in the same place. Sometimes we realize that we're in the wrong place. And sometimes we need to move because the world moves around us.

If you live in a major North American city on a lake or the ocean, there's a good chance that you've heard warnings that some beaches are closed or unsafe for higher-risk visitors because of high bacteria levels, typically after a major storm passes through. I grew up with these sorts of warnings for Lake Ontario. They're the result of a historic accident of timing more than a century ago.

Back in the middle of the nineteenth century, when Joseph Bazalgette and his team of civil engineers came up with a plan to build out a sewer system under London, buildings commonly had cesspits to collect waste. They were emptied when they threatened to overflow and the waste could be put to good use as fertilizer or used in industrial processes, but sometimes it just ended up in the Thames. Whether it was in the streets and gutters or dumped in smaller tributaries, waste from the city eventually drained down to the marshy foreshore and then into the river, helped along by the rainy weather. Not surprisingly, disease was rampant, with cholera outbreaks a regular occurrence. But the widespread understanding at the time was that disease was linked not to the waste itself but to the foul smell that came along with it. Sickness was a result of breathing "bad air" (this idea lives on in the name for mosquito-borne "mal aria"). Avoiding disease meant avoiding this air by being on high ground or away from the densely populated city entirely.

Bazalgette's proposal incorporated a series of sewers to collect wastewater from the streets and cesspools of London, channeling everything downriver to the Thames estuary where the incoming and outgoing tide could sweep the waste off to sea. It was (please pardon the anachronism) a shovel ready project. Then, for a few weeks in 1858, weather conditions meant that the reek of the polluted river was so much worse than usual that the event became known as the Great Stink. At the Houses of Parliament, right on the riverbank in the center of the city, it was intolerable. Legislators signed off on

Bazalgette's funding and construction started almost immediately, with London's first municipal sewer system largely complete by 1870.

Just a few years prior to the Great Stink, in 1854, Dr. John Snow was treating patients in the Soho neighborhood during a cholera outbreak. By carefully mapping the locations of homes in which deaths occurred, he was able to show that they were clustered around specific water pumps—strong evidence that the cause of the disease was contaminated water, not miasmic air, and incidentally birthing the science of epidemiology. The evidence was compelling enough that Snow famously carried out a public health intervention by arranging for the handle of the contaminated pump to be removed so local residents couldn't drink from it. But it wasn't until after an 1866 cholera outbreak that the main proponents of the miasma theory were fully convinced that the disease was waterborne. Today, of course, managing waterborne pathogens is an important element of sanitation systems. Cholera has never gone away. It's merely held in abeyance, and when these systems fail, it can come back with a vengeance. That's what happened in Haiti after the devastating 2010 earthquake that killed over two hundred thousand people. The lack of adequate water and sewage treatment in the aftermath made it possible for the pathogen to spread widely and become one of the worst cholera outbreaks in modern history, with more than half a million cases and over eight thousand deaths.

There's a sarcastic phrase among civil and chemical engineers—"the solution to pollution is dilution." Once a catchy summation of an engineering strategy, it's become a bleak joke because, in the twenty-first century, we know it's manifestly untrue. We have a better scientific understanding of the impacts of even low levels of pollution, and we're dealing with the accumulated waste matter from several centuries of industrial-scale production. (Most obviously, greenhouse gases that are diluted into the entire atmosphere of the planet have become the *problem*.) The solution to pollution is remediation or, preferably, redesign to avoid the pollution in the first place. But when Bazalgette was designing London's sewers, the conception of how disease was spread by invisible pathogens wasn't yet settled science, and dilution was welcomed as a way to manage the stink. By the time the germ theory of disease was fully established, Bazalgette's sewers and embankments had been under construction for most of a decade, with just a few years of work left before they were complete.

Mid-nineteenth-century London produced all sorts of wet waste—human waste, horse manure, slaughterhouse remnants, industrial waste—much of which ended up in the gutters. Rain, particularly heavy rain, was welcomed for the way it sluiced the streets clean of filth, diluting it in the process. There was no real reason to separate out what we now call sanitary sewers, which carry waste from our homes, and the stormwater sewers that carry runoff from the street. Other cities, including major cities in North America, followed the British example and built out combined sewage systems.

With a new understanding of disease came a new strategy: rather than just dilution, wastewater could be treated to destroy disease-causing bacteria. But when there's heavy rain, the volume of combined sewage can overwhelm the capacity of treatment plants. Even though the stormwater dilutes the sanitary sewage to the point where it seems clean to our nose and eyes, it can still be contaminated with dangerous levels of pathogens like *E. coli*. It's why beaches are closed for a day or so after a rainstorm, until the bacteria have had a chance to dilute further or degrade in the open water. The sewer systems of many major North American cities still echo these decisions that were made three thousand or more miles away and a century and a half earlier, before the germ theory of disease was fully understood. Sewage separation projects are now underway in many cities. Toronto recently built out a system to carry runoff from the downtown core through a stormwater-specific remediation system and then out to Lake Ontario. In other cities, the work is piecemeal: as the sewage lines come up for maintenance and repair, the combined sewers are being separated into sanitary and storm sewers.

Sewage systems around the world "locked in" a specific engineering approach to wastewater management and treatment, but new scientific understandings of infectious disease coupled with higher standards for what's considered acceptable meant that this approach has continued to evolve. The Deer Island Wastewater Treatment Plant that I visited had a whole new system added to it not long after it was built to meet the new, higher standards that resulted from the formation of the Environmental Protection Agency in 1970 and the passage and amendment of the Clean Water Act in 1972. There has never been such a thing as an absolute definition of "clean" water. We redefine what "clean" means on a regular basis as our ability to both measure contaminants and understand their health effects grows. The modern understanding among civil engineers is that, no matter what advertisements for bottled water might say, there is *no* water on earth that is pure or untouched. As my colleague Alison Wood, who researches wastewater systems, puts it: "all water is reused water."

But even without changing our standards for what constitutes clean water, treatment systems can become inadequate if they are allowed to degrade or if the quality of the source water is changed. There are thousands of American communities with water supplies that aren't considered to be fully safe to drink, most of them because their water sources have become contaminated, typically by local industry, and the municipality has neither the resources to bring the water treatment system up to standard nor the leverage to pressure companies to remediate the water sources. In Canada, dozens of First Nations communities have had drinking water advisories and boil-water orders in place for a year or longer, for a range of reasons: heavy metal contamination from industry, reactions with treatment chemicals that create unsafe water, or simply because their treatment plants weren't upgraded or replaced as they reached the end of their planned lifetimes or as the population they served grew. None of these are the result of physical threats, or technical threats, or anything that would be part of an action movie. It's just the result of regulations, funding, and oversight (or lack of oversight). But the end result—a community that doesn't have drinkable water—is functionally indistinguishable.

Finally, some contaminants are new in the world. On that field trip to our local water treatment plant in Needham, we asked the staffers what concerns were on their radar. Their response: "PFAS." Per- and polyfluoroalkyl substances are a group of a thousand or more synthetic chemicals that first began to be used in industrial and consumer products in the 1940s but whose ecological and biological effects became a matter of widespread concern in the early 2000s. PFAS are sometimes known as "forever chemicals" because they are stable and long-lasting—Teflon is an example—which means it's likely that they can accumulate in the environment and in body tissue. PFAS have mostly been manufactured and sold by only two companies, first DuPont and now 3M, but they are widely used and now found in drinking water and other sources. New regulations are coming into place and—for communities that have the resources to do so—new systems are already being retrofitted into water treatment plants to minimize any potential long-term health impacts.

The risks associated with maintenance and operation might be the hardest to address, but they might also be the most important. When people say that infrastructure is boring, this is what I think they mean. The effects of maintenance and renewal are invisible—if it's done right, nothing happens. But it's always a line item on a budget somewhere as an expense, and it's a tempting target to cut back. Like the effect of hungry termites, the effects of inadequate maintenance are invisible until they aren't. But the problem is not even that underinvestment in maintenance is what causes the failure. It's that undermaintained systems become more vulnerable to anything else going wrong—they fail when the predictable degradation over time meets an atypical challenge. Many of the biggest infrastructural failures that have happened in the United States in the past decades have been closely related to underinvestment in routine maintenance, including the California wildfires, the blackouts in Puerto Rico, and the multiday Texas winter blackout. And the undermaintenance of infrastructure doesn't occur in a vacuum, because it can't be isolated from design and planning, nor from renewal, rebuilding, and replacement. Insufficient maintenance can really be seen as a symptom of systematic underinvestment of societal resources in infrastructure.

The More Critical, the Less Appreciated

For those of us fortunate enough to have access to well-designed, well-maintained public utilities that mostly *just work*—they're there when we need them, twenty-four hours a day, making all the systems that we build on top of them function without fail—then it means that, by design, we can take them for granted. There's a human cognitive bias that the less something fails, the less we notice, appreciate, or value it, no matter how important it is to our lives. No matter who you are or where you live, anywhere on the planet, your daily life is either mostly about engaging in caregiving labor or made possible by caregiving labor, or both—by cooking, cleaning, hygiene, childcare, eldercare. As with infrastructural systems, it's when that labor is absent that it becomes most visible. During the peak of the coronavirus pandemic, in 2020 and early 2021, the U.S. (together with much of the world) was under a safer-at-home advisory, which meant that households were largely responsible for taking care of their own daily needs, including childcare, cleaning, and meal preparation, with little outside assistance. It brought this caregiving work into sharp focus for households, especially for those who had previously been able to outsource these daily labors. Alongside this, there was a new understanding of who the "essential workers" in our society are, including those in health care, food production and distribution, and utilities operation and maintenance.

What defines infrastructure is precisely that we don't have to think about it. It's not invisible so much as it's transparent—we see through it to the tasks that it enables. The idea of infrastructure as public utilities, and as a political right, implies that it's available to everyone in ways that are largely independent of economic status or geography—that everyone should be equally well served by water and electricity and telecommunications. And as long as these systems are running well, we can mostly ignore them, even as we depend on them. Being part of a community that's embedded in and enabled by infrastructural networks, particularly ones that meet survival needs like water and energy, means that we can outsource their operations and oversight to a relatively small number of specialized professionals. We might all need these systems, but working to meet these needs is no longer daily survival labor for *anyone*, much less everyone. In *Walden*, Henry David Thoreau wrote, "A man is rich in proportion to the number of things he can let alone." Not thinking about infrastructure, ever, is an indication that it's doing its job of invisibly supporting the rest of our lives. Infrastructural systems themselves are a form of care, made possible both by technologies and by the daily work of humans. But just as we still want to recognize, appreciate, and value the human caregiving labor that makes our lives possible, we need to do that for infrastructural systems too, including the people whose skill and labor keeps them functioning.

The journalist Michael Lewis described something that he calls "the fifth risk" in his book of the same name: it's the risk a society runs when it falls into the habit of responding to long-term risks with short-term solutions. What I've been describing here—the sort of underinvestment in infrastructure that leads to poor grades on an infrastructure report card—is part of that fifth risk. What societies need instead, argues Lewis, is "effective program management": all the actual boring work that needs to happen in the technologically complex world we live in. Building new facilities, with the exciting ribbon-cutting ceremonies that come with them, is a part of this work, but mostly it's having a checklist and a schedule and a cohort of trained staff who can carry out routine inspections and maintenance of all the bridges in a region, with all of the funding and policy that entails, in a program that's stable over a span of decades or longer. It also means careful regulation and adequate oversight of infrastructure providers.

Like many other smart, technically minded, and not terribly socially well-adjusted adolescents, I read Ayn Rand's novels *The Fountainhead* and *Atlas Shrugged*. The protagonists include an architect, an engineer, and a railway magnate, and the novels are an extended dramatization of Rand's philosophy of objectivism, a sort of libertarianism that valorizes making new things and also money, with a corresponding deprecation of any work that has to happen again and again, every day. I read them while I was in engineering school, and I remember deciding at the time that it was pointless to make my bed and so I wasn't going to bother, because I was going to be an *engineer* and a *scientist* and I was going to bring new things into the world. By the time I was in my early twenties, my adolescent infatuation with the books had passed, helped along by a few things: by how anti-feminist they were (I wanted to be the male protagonists, not the female characters who were largely subservient to them); by going to graduate school to investigate biological materials and learning firsthand the reality that much of science progresses by careful, repetitive daily work; by having been responsible for my own household and its daily chores for some years—cooking and cleaning and brushing my teeth, and even making my bed some days—but mostly just by growing up enough to start to fully *see* all the people around me, and the roles that we play in each other's lives. But it's still really clear to me that many of Rand's ideas are prevalent in culture, especially American culture, and especially in engineering and technology: they prioritize making things over caregiving and maintenance, the new and novel over what's sustained, ideas and artifacts that can deliver a profit over activities that need an ongoing commitment of money and other resources, and individualism over acting collectively. As an engineering professor, I see examples of this nearly every day. It's all about "Design new things and change the world!" Don't get me wrong—I like new things, all the fruits of the technological development that built the world we live in, as much as the next person. But our sustained and sustaining systems are widely considered unglamorous and boring.

There is no efficiency in childcare. There is very little scalability in nursing. Dental cleaning or cardiac surgery or pulling espresso shots or replacing tires takes about the same amount of a skilled worker's time, year on year. In the classic example, the time it takes for a string quartet to prepare for a recital hasn't gotten any shorter in the past century, even as new technologies and modes of production have increased productivity and lowered costs in the world around them. Economists call this discrepancy "Baumol's cost disease," and even the name itself is a tell, since calling it a disease categorizes it as pathological, something to be cured. But these kinds of skilled human labor, especially caregiving, can't be made to happen faster, or at scale, and so they do not get cheaper with time. Nor should they. A former student of mine, Mikell Taylor, is a roboticist, and she tells the story of hearing one of her professors, my former colleague Gill Pratt, respond to a reporter who asked him about designing robots to take care of his children so he could concentrate on his work. Gill responded that the reporter had it exactly backward—that he wanted to design robots to do his work so he could spend more time with his children.

It's easy to underappreciate the caregiving, whether human or technological, that makes our lives possible as long as it's working well (arguably, in fact, *because* it's working well). All of this makes it hard to understand the value of doing the work, and makes it easy to make decisions to scale back spending, since the best-case scenario for steady, reliable infrastructural systems, whether public or private, is that no one notices them. The replacement cost of the Genoa and the Minneapolis bridges were each in the neighborhood of a quarter of a billion dollars. There's a strong incentive for large contracting firms to bid on projects like these, and to pay lobbyists to help make sure they get funded and built. In contrast, the financial benefits of routine inspection, maintenance, and renewal are diffuse—anyone directly involved might make a steady salary, but no one is likely to make a windfall profit. That societal bias toward making and novelty and away from caregiving and maintenance is reflected in who makes money and how.

What's more, the nonmonetary externalities of the careful maintenance that keeps infrastructure functional are even more diffusely distributed. Both the Genoa and Minneapolis bridges were crucial transportation links in their regions—it's why they were both immediately rebuilt—and benefited hundreds of thousands of people. Every power outage, however short, exacts a cost in time and agency from those affected. So does every water main break and boil-water order. Having robust, reliable infrastructure doesn't just happen. It's a collective form of care, paid for collectively, that creates both shared and individual benefits. If one important role of infrastructure is to provide for basic needs in order to enable a higher degree of agency, then universal provision has the capacity to decentralize power in a community. Conversely, when infrastructural systems begin to degrade and fail, the negative impacts are widespread. It's those diffuse, nonmonetary externalities that really make it clear that while these infrastructural systems are obviously technological, they are also deeply social systems, embedded in daily lives, in the economy, and in political structures. It's why utilities, whether private or public but especially network monopolies, need to be carefully regulated to balance that intrinsic concentration of power. It's why there is a relationship between well-maintained systems and effective governance, because this governance is how we make that diffuse power actionable and allow for collective decision-making. And it's why seeing, recognizing, and valuing these systems is important. But increasingly, it's not going to be enough.

The Oncoming Threat to Infrastructural Systems

During the California fire season, crews go all over the state to fight wildfires. We're used to thinking of firefighters as "first responders," but the first people who go out are California Department of Transportation (CalDOT) workers. They inspect the fire-ravaged roads to make sure that they're safe for the firefighting crews and all their gear, and make emergency repairs if they aren't. Utility workers are the "zeroth responders." In the U.S., cooperative systems among public and private power utilities help to make sure that lineworkers are available to make emergency repairs necessitated by severe weather. Crews from northern regions head south to help deal with the aftermath of hurricanes, and workers from warmer southern regions are staged across the northern parts of the country when a blizzard is bearing down. The power outages in Puerto Rico after Hurricane Maria were so widespread and long-lasting because the grid was already fragile as a result of a chronic lack of investment and maintenance—further to the point earlier about the role of equity and governance in infrastructural systems—but also because there was no plan in place to bring in lineworkers via these networks of mutual aid. Well-maintained systems with sufficient investment, funding, and staffing are more resilient to challenging environmental conditions, not just storms but also extreme heat or cold, or floods, or drought. But this is now basically table stakes when it comes to severe weather, because what we're playing for has begun to get much, much higher.

Solar storms like the Carrington Event are a huge risk, but though their exact timing is unpredictable, we still know that they're coming and can prepare. Neglected maintenance of infrastructural systems are the red termites—the widespread but underappreciated threat that is undermining their foundations. But climate change is the massive, obvious risk to nearly all our infrastructural systems and the ways of life that they enable. Extreme weather events have begun happening, and are going to happen a lot more as the greenhouse gases in our atmosphere lead to climate change. It's no longer enough to do a good job of appreciating, maintaining, and staffing the systems we already have, because the stable world that they are built for—the underlying landscapes that our infrastructural systems are designed to traverse—has already begun to disappear.

Eight

INFRASTRUCTURE AND
CLIMATE INSTABILITY

Over the last two decades, I've watched the South Bank of the Thames boom, becoming home to skyscrapers like the Shard, the tallest building in the UK. When I was in London Bridge station, next door, while it was being rebuilt, our guide, a representative of the lead contractor on the project, casually mentioned that they were required to guarantee their work for a hundred years. A *hundred years!* Then I remembered that the station had opened in the 1800s—in 1836, it turned out, making it among the oldest railway stations in the world. A century-long commitment for a station that's already nearing its bicentennial suddenly seemed a lot more reasonable.

Stability of all sorts, including political and economic stability, is what makes it feasible to front-load large investments of resources with the expectation of continuing benefits over a long period. What we think of as the primary characteristic of infrastructure—that it just works, and works invisibly, to support other activities—is made possible, in part, by that stability. Over the history of the station, trains went from being powered by coal to being powered by diesel to having electric locomotives. Vehicles became faster and more powerful, expanding and reshaping the transportation networks of the entire region. But what didn't change much over the first century and a half of the train station's lifetime was London's climate.

Our current geological epoch, the Holocene, began with the end of the last ice age, about twelve thousand years ago. Since then, our climate, landscape, and sea level have been largely stable. It's this stability, in fact, that's thought to have made the Neolithic Revolution possible, including the rise of agriculture, human settlement, and what we recognize as civilization. This stability also underpins our infrastructural systems since these networks traverse the landscape, embedded in it and transporting resources across it. Until very recently, humans have been creating infrastructural networks with this implicit understanding: the weather might be variable, but the underlying climate would remain stable.

That expectation has driven how we think about and build infrastructure. It suggests that well-designed, carefully maintained, and thoughtfully remodeled systems could remain robust and functional over decades, even centuries. Stability of all sorts, including political and economic, makes it feasible to front-load large investments of resources with the expectation of continuing benefits over a long period of time. What we think of as the primary characteristic of infrastructure—that it just works, and works invisibly, to support other activities—is made possible by stability. A habitable planet, with a stable climate, is the infrastructure to our human-made infrastructure.

If our metaphors for risk include red termites, chewing away the underfloor beneath us in a crisis of inadequate care, or a gray swan that can suddenly come flapping in through an open window in the form of a major earthquake or the next Carrington Event, then climate change is, in policy researcher Michele Wucker's memorable formulation, a gray rhino. Like the proverbial elephant in the room, a gray rhino is a high probability, high impact threat, but one that needs to be seen and discussed rather than ignored, so we can figure out how to deal with something that big that's already in our living room, knocking over the furniture and leaving gouges in the floor.

In the nineteenth century, global industrialized energy usage in the form of fossil fuels began growing exponentially, doubling roughly every two decades. In 1856, Eunice Foote filled a glass cylinder with carbon dioxide and another with air, put them both in the sun, and measured their temperature. She became the first scientist to demonstrate that carbon dioxide is a greenhouse gas. By the middle of the twentieth century, the threat of global warming as the result of greenhouse gas emissions was recognized and articulated. In 1880, the estimated concentration of carbon dioxide in the atmosphere was about 290 parts per million; in 2019, the global mean was 415 ppm. Because it scales with fossil fuel combustion, the concentration has been increasing more and more steeply. Of the ten warmest years globally in a century and a half of reliable record-keeping, nine of them were between 2013 and 2021, the most recent as of this writing. A familiar visualization represents the average global temperature for each year in colored stripes, mostly blue and pink in the past but increasingly red and darker red towards the present. Anthropogenic climate change is here. What does it mean for our infrastructural systems?

Infrastructure in a Warming World

The first-order effect of climate change is warmer average temperatures, which means that the hottest days will be hotter and heatwaves will last longer. Much of the population growth in the southern U.S., especially the southwest, was shaped by access to air-conditioning and electricity to power it, the demand for which will increase with higher temperatures. At the same time, places that have historically been cooler in the summer, like the Northeast and the Pacific Northwest, are getting hotter. Here in Boston the temperature usually falls to a comfortable level at night even during July and August. But hotter days, especially consecutive ones, mean warmer nights, which leads to keeping the air conditioner running rather than opening windows. That increased use of air-conditioning at night means power plants need to provide higher output overnight. All of this means seasonal demand on the electrical grid, which in the U.S. peaks in the summer (as of this writing, the historic peak in the afternoon during a countrywide heatwave was in July 2019), is set to keep increasing; without a commensurate investment in supply and transmission, it means more days at the edges of its capacity.

Transportation networks are also affected by generally hotter weather. In places that have historically been more temperate, transportation infrastructure like roads, bridges, and railways may not be designed for higher temperatures. Rails, for example, expand in the heat and can buckle. Hot days might mean that trains need to run at slower speeds, or that service is suspended entirely, as during unprecedented hot weather in Portland, Oregon, in 2021, and then in the UK during the continent-wide heatwaves of 2022, including the August day when temperatures in the country reached 40 degrees Celsius for the first time on record. That European heatwave affected other infrastructure that relied on cooling—some data centers in the UK were shut down, and French nuclear plants obtained special waivers to discharge warmer-than-usual cooling water because the intake water was so much warmer than usual.

In many regions, a warmer climate will also mean a drier landscape, either because warmer temperatures lead to more evaporation or because of changes in the amount and timing of precipitation. Hotter, drier summers means higher wildfire risk, as fires catch and spread more easily and burn more intensely in dry vegetation. In some regions, there's an increasing risk of drought, shading into multiyear droughts, shading into desertification.

In the U.S., the hundredth meridian has historically been considered the *de facto* dividing line between the semiarid climate of the West and the more humid continental climate of the East. The transition is easily visible from the air because it's where fields change shape, rectangular on one side and circular on the other. Industrialized agriculture west of the line depends on augmenting the limited rainfall, most commonly with center-pivot irrigation. A quarter-mile long sprinkler arm, supported by wheels along its length, rotates slowly around a point and sweeps out a circle in which crops are grown. The water mostly comes from underground aquifers, and the build-out of the electrical grid across the continent, which made it possible to pump that water to the surface inexpensively and at scale, transformed land use in the region. Unlike surface reservoirs that are refilled seasonally, the aquifers recharge extremely slowly, so slowly that they're often described as reservoirs of "fossil water." It's not sustainable, and after a century or so of electrically powered pumping, the water level in the region's aquifers, including the massive Ogallala, has dropped precipitously. Less and less water is accessible for use, including by the communities that depend on the same aquifers for daily needs. A warming climate in this region would mean more irrigation over a larger area. Already the dividing line sits closer to 98 degrees west longitude than 100, and climatic models suggest that it will continue to move eastward, driven by faster evaporation from a warming landscape. Without a transition of industrial agriculture to practices that re-quire less water or are more drought tolerant, this may mean abandoning growing crops in the region altogether.

Farther west, California's Central Valley is the growing region for nearly all of the country's almonds, pistachios, walnuts, figs and dates, grapes for raisins, and plums for prunes, as well as for vast amounts of familiar super-market produce like lettuce and berries. The fields of the Central Valley are so huge that even at seventy-five miles per hour, the minutes tick by as you drive past them. They're made possible by irrigation from California's mas-sive, interpenetrating water systems. Much of that water starts off as winter snowpack in the Sierra Nevada mountains, where warmer weather might mean more rain and less snow. That means less stored water, not just in the reservoirs filled by rushing meltwater in the spring but because the snowpack *itself* provides a huge amount of storage capacity. This in turn affects the availability of water not only for agriculture, but also for household use and hydroelectricity. In the western U.S., where regional systems span state and international borders, allocating water for these critical needs has never

been simple or harmonious and will only become more complex and contentious if less water is available to share. A more subtle challenge has to do
with timing. The accumulated snow in the mountains of northern California melts gradually over the spring and summer, which means that reservoirs are refilled and there is a steady supply to the water system through the
all-but-rainless late summer and early autumn, delivering water during peak
usage in the Central Valley as harvest time nears for the growing crops. But
a smaller snowpack in the mountains will also melt faster, meaning not only
that there will be less water available late in the season, but that peak water
delivery may no longer coincide with peak preharvest demand. The landscape and climate that it was designed for no longer exist.

This is obviously just the barest sketch of the ways that infrastructural
systems are affected by warming, and of the host of changes and adjustments—
if not outright reconsideration, redesign, and rebuilding—that will be required for them to continue to function effectively in a warmer climate. But
anthropogenic climate change means more than just a steady-state increase
in temperature.

The Cumulative Impact of Design Days

The engineers that spec out the heating and cooling systems for buildings
(known as HVAC, for "heating, ventilation, and air-conditioning") use the
term "design days" for the hottest or coldest days of the year, the days that
demand the most of building systems, and they are usually design days for
the underlying infrastructural utilities, too. In the metropolitan Boston area,
where most homes are heated with utility gas, demand during cold snaps in
the dead of winter can max out pipeline deliveries into the region. To ensure
that there's enough fuel available, liquefied natural gas is delivered by ship
to a facility in Boston Harbor where it's turned back into a gas and injected
into the distribution network. Delivering service at peak loads is often the

biggest challenge for infrastructure. It frequently requires pricy solutions, whether it's closing down a busy harbor to safely deliver tankers full of fuel or firing up expensive peaker plants to ensure there's enough electricity in the grid.

What's more, design days are *risky* days—when the entire capacity of a system is in use, there's little redundancy or room for error. In August 2020, a critical shortage of energy in California meant that an estimated four million people had their power shut off during a heat wave; a similar widespread blackout was feared on that hot afternoon in 2022 when the text message went out asking residents to conserve power. During the European heatwave in July 2022, blackouts in East London were narrowly averted by the purchase of electricity from Belgium at *fifty* times the normal cost, nearly five times the previous record high. It wasn't even that there wasn't enough electricity on the island of Great Britain; it was because of a bottlenecked electrical grid with no remaining capacity to transmit energy south from Scotland to the capital.

A few miles downstream of London Bridge is the Thames Barrier, a line of gleaming steel seashells that crosses the river, each a hydraulically powered gate that can be raised to protect the low-lying city from flooding due to heavy rains or storm surges. The barrier was designed in the early 1970s and became operational in 1982, with the expectation that it'd be used two or three times per year. In the 2000s, the gates were closed six or seven times each year; in 2013, the barriers were raised dozens of times due to heavy rains and to limit the tidal effects in the city. Current models suggest that the Thames Barrier can remain functional until 2070 or so, even if it's used more than a few times each year, although it will require something like £3 billion in maintenance work over that time so it can continue to be among the floodwater protection strategies in the climate adaptation plan for the London area.

More extreme weather conditions are part of the challenge of anthropogenic climate change. Civil engineers design infrastructural systems for bad weather by analyzing all the data that's available in the past. A "hundred-year storm" means there is a 1 percent chance that a storm with that severity will hit in any given year, so if you're building a facility or system that might last for decades, there's a reasonable chance that one of those storms will hit during its lifetime. It's worth designing critical utilities to withstand them, along with a healthy margin of error. But what if the climate of the future isn't like that of the past? What happens if hundred-year storms arrive more frequently, and the "new" hundred-year storms are even more severe?

In October 2018, for the first time in its history, the city of Austin, Texas, issued a boil-water order. Heavy storms and flooding (the third "hundred-year storm" in the region in five years) had washed silt and debris into the reservoir that supplies the municipal treatment plant, overwhelming the design capacity of the system. It wasn't a lapse in maintenance or a design failure—it was simply that the worst-case scenario of the present day had exceeded what the civil engineers had designed for.

These types of failures are set to become increasingly common worldwide. Even small, gradual changes in temperature, in average rainfall, or in sea level can have noticeable effects. For example, climate models predict that hurricanes will become more intense and damaging. While the specific dynamics are complex, the energy that powers hurricanes comes from thermal energy. Warmer air over the ocean holds more moisture to fall as torrential rain; warmer ocean water powers the hurricane-force winds that damage buildings and down power lines. Higher sea levels mean higher storm surges and therefore more coastal flooding. Even if they don't hit their design limits and avoid being fully overwhelmed, more severe weather means that infrastructural systems will be operating closer and closer to the edge of capacity and are likely to need more maintenance to keep them in good repair.

From Global Warming to Global Weirding

Engineers and scientists have a term for when someone gets really, really upset at something that seems inconsequential—they call it "going nonlinear." In mathematical terms, a linear response is proportional to the input: a gentle stimulus evokes a mild response, a stronger stimulus evokes a more robust output. When something—or someone—goes nonlinear, all bets are off. The response might be all out of proportion to what happened, like someone getting toweringly angry in response to a minor quibble.

As our planet gets warmer, familiar events like heatwaves will increase in severity or duration, with largely predictable effects on our infrastructural systems. But there will also be more events that are nonlinear—harder to predict, and with much more severe effects. Giant wildfires producing world-girdling smoke on both sides of the Pacific Ocean in 2020 (Australia's "Black Summer" at the start of the year was followed by a record-setting California fire season at its end) is a decidedly nonlinear response to a small increase in temperature for a few months and the somewhat drier vegetation that resulted. This is not so much global warming as global *weirding*. All of our infrastructural systems—for that matter, nearly all human systems—have been created assuming that, however variable the weather might be, the underlying climate is stable. Now we are starting to deal with the very real possibility of human systems being out of spec because, for the first time in all of human history, the climate is changing rapidly and potentially in ways that aren't easy to model or predict, or that we don't yet have the data to understand fully. This includes the possibility of tipping points. What might happen if the Arctic permafrost melts and releases trapped methane, itself a greenhouse gas? Or if ocean circulation systems, like the North Atlantic Drift, weaken or even collapse? Without currents which transport warm water to higher latitudes, western Europe might become as cold as western Canada.

These are all potential effects of climate change, in the near future and with modest amounts of warming. But what really matters is not so much the effects as the *impacts*: food and water insecurity, the danger of extreme heat, especially for the large fraction of the world that can't just turn the air-conditioning up, the likelihood of large-scale climate migration, and the human suffering that will result from all of the above, either directly or as a result of resource conflict (to say nothing of the toll on nonhuman species and the likelihood of irreversible damage to ecosystems).

In a world where climate stability is being replaced with instability, infrastructural systems are simultaneously more vulnerable and more necessary. The degree to which these systems remain functional is often exactly what determines the severity of a disaster, whether it's the roads and other transportation networks that make it possible to evacuate a city in the face of wildfire threats, or the water systems and electrical grids and telecommunications networks that stay operational when a hurricane makes landfall and allow residents to safely shelter in place. There is no response to climate change that doesn't involve infrastructural systems.

But it's not just that our infrastructural systems will be—are being—affected by anthropogenic climate change. The real problem is that our infrastructural systems consume energy, and as long as they are primarily powered by fossil fuels, they will continue to be one of the main *drivers* of climate change.

Collective Systems Mean Collective Energy Use

To do anything—to actively engage in the world in any way—requires energy. Energy that comes from fossil fuels results in greenhouse gas emissions and as long as access to energy means combustion, then energy and carbon footprints will be closely related, if not outright synonymous.

In the U.S., about half of the pollutants that contribute to anthropogenic climate change are directly attributable either to the energy and electricity we use in our homes and workplaces or to vehicles on transportation networks. The U.S. isn't an outlier—this pattern of energy usage and emissions is pretty typical of industrially developed countries in the Global North. The majority of greenhouse gas emissions are mediated through the shared technological systems of infrastructure. Anthropogenic climate change poses a threat to infrastructural systems, but our infrastructural systems are also the biggest *contributor* to climate change.

That means we find ourselves in what seems, on the surface, to be an intractable conundrum. We've created these collective infrastructural systems that make our lives, as we know them, possible. Any future with limited, reduced, or even more frequently interrupted access to them is recognizably worse than our present, if not outright dystopian. We can't stop using them.

But by their nature, infrastructural systems are particularly vulnerable to climate instability. The worse it gets, the less well these systems will work. And because the energy that powers infrastructural systems largely comes from fossil fuels, the more we use them, the worse it gets.

But there's a way out. Or rather, a way *through*.

The first step is to break the connection between combustion and energy, to make it possible to generate, distribute, store, and consume energy without producing pollutants. As a civilization, we've spent most of the past century developing a way to do that: electrification, and working out how to produce light, heat, and motion, on scales ranging from the tiniest LED to the largest industrial facilities. Hydroelectric and nuclear power plants are large-scale, twentieth-century technologies for harnessing energy without combustion; facilities like Dinorwig are a means of storing that energy at scale; continent-wide electrical grids allow for distribution. They've now been joined by twenty-first century technologies, those invented, developed, and commodified over the past several decades: wind turbines, solar panels, electric vehicles, grid-scale batteries, even the microwave oven, and the communication and computation infrastructure that allows for the coordination of all these systems.

As long as the only way to access appreciable amounts of energy for human use was through combustion, the only way to reduce emissions was by using less energy. By transitioning our technological systems to instead use electricity, generated from renewable sources, we can keep all of the social and economic benefits of access to energy, including all of the agency that it provides and the ways of life that it supports, without paying the self-defeating (or outright ecocidal) cost of putting more carbon dioxide into the atmosphere.

In a very real sense, anthropogenic climate change has been happening for the past two centuries, since the beginning of that exponential growth in energy usage powered first by coal and then by petroleum and gas. That elapsed time means that there's been enough accumulation of greenhouse gases in the atmosphere and enough time to collect (and learn how to collect) data that we can experimentally observe its impact. Because we've moved further up an ever-steepening exponential growth curve, the *rate* of change is much higher. While there's no "acceptable" amount of global warming, there's also no point at which all is lost, no point at which it is futile to try to limit emissions. A world with 1.5 Celsius degrees of warming means less suffering and damage than one at 1.6 degrees, which is less than at 1.7 degrees. Reducing or eliminating emissions will limit warming and make the resultant climate instability easier to address. If we want to keep infrastructural systems functioning in a world of anthropogenic climate change, mitigation is key. Limiting the amount of carbon dioxide in the atmosphere means that we can begin to stabilize the climate and give ourselves more breathing space to respond, rather than locking ourselves into an exponentially worsening situation. It's like having an overflowing bathtub: yes, you absolutely need to start mopping up the floor to keep the water from getting into the walls and causing permanent damage, but your first priority always has to be turning off the faucet.

Transitioning all of our infrastructural systems away from combustion is a transformative undertaking. It means replacing and rebuilding systems at all scales, from the domestic to the global, everything from backyard grills to container ships. It's daunting, but we can incorporate everything we've learned over the past century and a half about how these systems function and use all the new technologies that have become available. Our responses to climate change don't have to be purely reactive, focused on averting catastrophe. Instead, we can act on the opportunity to build a sustainable future of thriving human communities embedded in healthy environments.

If infrastructure is the biggest lever we have to mitigate and respond to climate change, it's also a lever that none of us can move alone. Infrastructural systems are collective ways of providing us with individual agency, and this works in the other direction too—our individual energy usage, and thus our individual emissions, largely result from these shared systems, which serve hundreds, thousands, or even millions of other people and households. Infrastructural systems lock in efficiencies—of energy, but also money and time—through networks and standards, and that means they also lock in specific ways of doing things, making them harder to do in other ways. That's what makes it possible for us not to have to think about them, but it's also what makes it challenging—often prohibitively so—to act in ways that aren't the ones supported by these systems, like not driving a car in a city with limited public transit options. What's worse, we know that making a personal decision, like deciding to leave the car at home or replace the gas furnace in our house with a heat pump, doesn't necessarily help anyone else make the same decision, and because these are collective systems, our individual contribution is minuscule. Many of us have come to the disheartening conclusion that, no matter how great our individual sacrifice of time, money, or energy, no matter how much we radically change our own lives and homes in order to reduce our personal emissions, we can't move the needle alone.

The Limits of Individual Action

Every member of my immediate family lives on the other side of a national border, or an ocean away, or a continent away. As an academic, I've lived, worked, and traveled around the world, with friends and colleagues and collaborators in many countries. Air travel in the U.S. is relatively inexpensive, and flying has been important to me for my entire life, personally and professionally, starting with those flights I took with my family when I was too young to even remember them. If I think about greenhouse gas emissions from an individual perspective, I know that flights are going to be a big contributor for me. How big, exactly? To find out, I started at the Carbon Footprint website, which has a calculator to figure out your emissions footprint so you can purchase the appropriate carbon offsets in order to be a net zero producer of greenhouse gases.

For the last decade or so, I've driven a car with a conventional gas engine that is compact and reasonably fuel efficient, a 2013 Mazda3 hatchback. I plugged its make and model into the emissions calculator, ballparking my total annual mileage as twelve thousand miles, about average for American drivers. The calculator spits out that my car produces about 3.12 metric tons of carbon dioxide each year. Easy.

I have family in Los Angeles—my brother, sister-in-law, and my beloved young niece and nephew. What are the greenhouse gas emissions associated with flying out to visit them? I enter the Boston and Los Angeles airport codes, choose "economy" (business and first-class passengers, because there are fewer passengers in the same space, are effectively responsible for a higher proportion of emissions). I specify a return trip, since I'll come back to Boston after my always-too-short visit. Finally, I click on a checkbox next to the option to include radiative forcing, to account for the way that combustion products from jet planes have a disproportionate effect on global warming because they're released high in the atmosphere. The calculator estimates my effective emissions from a flight out to Los Angeles and then home again: 1.17 metric tons of CO_2. That *one* cross-country trip—one visit with my niece and nephew—produces over a third of the emissions that my car does. It's the equivalent of four months of driving back and forth to work, running errands, driving out of the city to visit friends, everything that I do in my car. To be fair, a coast-to-coast return trip is also more than five thousand miles, nearly half of my annual driving mileage. Jet planes go through a lot of fuel, but those emissions are shared by everyone on board, and so the emissions per-passenger-mile are slightly lower than when I'm alone in my car. Yet even with a daily driving commute, most of my personal contribution still comes from flying. It only takes a couple of flights to LA and one or two home to see my parents and my sister's family, outside Toronto, for me to double my personal carbon footprint, and that's just for my family—never mind seeing my friends or work-related travel.

It's certainly possible for me to cut air travel out of my life, which would significantly reduce my individual carbon footprint, if at the price of only rarely getting to spend time with many of the people I love. But here's the thing: even if I did stop flying, if *everyone* stopped flying, it's not going to make much of a difference. It turns out that all air travel in total only accounts for a few percent of total greenhouse gas emissions. The contribution of civil aviation is dwarfed by ground transportation and power generation. This is a perfect example of why making individual changes feels so demoralizing: if I decided to radically reduce my personal carbon footprint by never flying, at tremendous personal cost, it wouldn't change anything else in any appreciable way.

We now know this empirically, because many of us were part of a global natural experiment on the effect we can have on climate change by dramatically curtailing our personal mobility. In March 2020, the World Health Organization declared a global pandemic, and within days Massachusetts and much of the rest of the country—and a large proportion of the world—had a crash-stop of almost all voluntary, individual use of transportation. I drove so little over the following few months that my car's brake pads rusted out and needed to be replaced, and my mechanic told me that they'd seen many cars like mine. Closed borders, travel restrictions, and mandatory quarantine periods put an end to all but the most necessary air travel. I didn't get on a plane again until June 2021—by far the longest stretch in my adult life that I didn't fly. For over a year, the two biggest contributors to my individual greenhouse gas emissions were almost entirely eliminated. I radically reduced my personal carbon footprint, just like I was supposed to. So did millions of other people.

The economic cost of the COVID-19 pandemic was enormous, but the personal, social, and emotional toll that it took on human connections continues to be incalculable. All the people who didn't get to be with their loved ones for a year or more, or when they were ill or dying, or ever again. It was the most catastrophic and traumatizing way imaginable of reducing our level of agency around mobility and transportation, but it did have an effect: greenhouse gas emissions in the U.S. dropped sharply in 2020, by around 10 percent. All of those hugely changed behaviors, and the total effect was only about ten percent. Globally, the drop was even smaller, about 5 percent. And that decline didn't last. After rebounding in 2021, global emissions are back to their pre-pandemic trajectory.

What this showed us, powerfully, is that individual, voluntary actions aren't the main driver of our collective greenhouse gas emissions. Most of our total energy usage, and therefore the bulk of our emissions, is collective— industrial usage, food production, energy generation, and goods transportation. None of these are tractable to individual decisions about reducing our carbon footprint.

So where did the idea come from, and why is it so powerful? The genesis of the "carbon footprint" was an early 2000s advertising campaign for British Petroleum. Rather than making carbon emissions the responsibility of oil producers, or even regulators, this rhetorical move shifted responsibility to consumers—to individuals, in other words. After all, what responsible, moral person who cares about the planet and wants to do something about climate change *wouldn't* want to reduce their emissions? But it's impossible to create deep or sustained reductions in emissions by changing our behaviors if we're not also changing the systems themselves. Which means that we, the consumers of energy, are being set up to fail and to feel bad about not being able to solve something that we can't possibly solve on our own. Worse, we're being set up to make *others* feel bad about not making "better choices." Our inability to solve the problem as individuals corrodes our belief that reducing emissions is even possible, and our motivation for trying: "What difference does it make if I fly less if everyone else is still going to conferences and off on vacation? And anyways, it's all a rounding error compared to cars and trucks." Or it corrodes our solidarity with others: "I'm doing my part and making a sacrifice, why aren't they?" It's a stellar example of the truism that what all advertising sells, regardless of the product on offer, is insecurity and discontent.

A "carbon footprint" implicitly conflates energy usage with using—that is, *buying*—fossil fuels, which means that reducing our individual carbon footprint requires reducing our *energy* footprint, which also means less of everything we could do with that energy. Conversely, if we want to continue benefiting from energy, or if we want those benefits to be extended to more people, that means burning more fossil fuels, even if it means a higher carbon footprint. Formulating it as an equivalence makes it very clear why oil companies would pay ad agencies to promote the idea of carbon footprints. Cementing the relationship between petroleum products, energy, and freedom is a good way to ensure continued demand for your product.

It's also a misdirection. A higher energy footprint only means a higher emissions footprint if that energy comes from combustion. But if it doesn't— if we power our infrastructural systems, particularly the electrical grid and transportation—with energy that *doesn't* come from fossil fuels, we can decouple our energy footprint and our carbon footprint. We couldn't have done this a generation ago. Back then, the only way to reduce emissions was to reduce energy usage, either directly or through increased efficiency (like how LED lightbulbs use much less energy to produce the same amount of light as incandescent bulbs). What's changed is the development and commodification of energy alternatives like solar power, wind turbines, and grid-scale storage, all of which augment "traditional" sources of renewable energy, including hydroelectricity and nuclear power. Renewable energy is now where electricity itself was a century and a half ago: the technologies are still somewhat emergent, with some unresolved questions, and they're still only a small part of energy systems. But the writing is on the wall: it's now just about possible to create infrastructural systems that are fully powered by sustainable and renewable energy, ones that produce and are powered by the same amount of energy, or even more, as their petroleum-burning predecessors, but without the associated atmospheric pollution.

This possibility allows us to see a future where we don't have to choose between a stable climate and all the agency and opportunity that infrastructural systems provide. We can rebuild infrastructural systems to make them not just resilient but sustainable, to do more with them rather than less. We have the unprecedented opportunity to extend these systems, and everything that they enable, to more people and communities. But we can only do it if we act collectively. It means shifting from the mindset of individuals acting inside, or even in direct opposition to, our collective systems, and instead acting together to transform them.

Toward an Energy Transformation

Year-over-year total electricity usage is fairly steady in the more industrially developed (OECD) countries—the demand is effectively saturated, in that residents of these countries might do different things with the power but it doesn't change aggregate electricity consumption by much. That's also been approximately true of energy usage broadly: increased wealth across a population leads to increased energy consumption, but only to a point. That flattening of the energy curve is also a result of a consistent, sustained effort to do more with less, including through policy and regulation like widespread efficiency standards, which cancel out increases due to economic growth. In non-OECD countries, the total electricity (and energy) consumption is growing as more and more people get access to electricity and the opportunity and resources to use it, and this growth is still largely fueled by fossil fuels with an associated increase in emissions. Reducing energy usage by doing less—not being with our friends and family, or otherwise reducing activities to use less power—is a tough sell. But to ask communities worldwide—both in the Global South and in underserved communities in otherwise wealthy, developed countries—to limit *their* energy usage, communities that may never have had access to these systems and the lives they enable, would be inequitable, bordering on unconscionable.

Our infrastructural networks, as designed, funnel resources to a fraction of humanity to allow them—and I'm part of this group, so I need to say "us"—to live lives of comfort and agency. They're the most powerful means that humanity has ever devised to take care of each other at scale, to use energy and resources collectively. These systems are how we reduce the quotidian struggle to be a human, to have a material human body with material needs in a material world. In a more equitable world, more people, globally, would have access to the benefits provided by shared infrastructural systems. But that also means taking a hard look at the inequitable distribution of the associated harms, and figuring out how to limit rather than displace them.

Infrastructural networks are also among our most powerful tools for facing climate change. We need them to both prepare for and respond to the growing impacts of a rapidly and unpredictably changing climate. The impact of crises like natural disasters and extreme weather events is entirely mediated by humans—who's in harm's way, what resources they have available, and the scale and effectiveness of the response. The tools that we use to respond require access to energy, materials, information, and the networks we use to move them all around—in other words, infrastructure.

At the same time, many infrastructural systems, as currently conceived and built, are just not going to work well anymore. Whether it's drawing down aquifers or using up fossil fuels while pumping carbon dioxide in the air, they rely on resources and processes that might not be viable a decade from now, much less a century, much less in perpetuity. They are, in the most literal sense, not sustainable. On top of that, these networks were built presupposing a climate that is already disappearing; and we don't really know for sure what it will change to, other than being much more variable and extreme. Even if all we care about is that the currently extant infrastructural networks—again, I have to use the first-person plural here, our networks—continue to function well enough that we don't have to think about them, we can no longer ignore their close relationship with anthropogenic climate change. Not the way they contribute to it because they're powered by fossil fuels, and not the way that being embedded in the landscape makes them unavoidably vulnerable.

Where does this leave us? For most of the two centuries that we've had modern infrastructure—infrastructural systems that use exogenous energy, mostly in the form of fossil fuel combustion—it was built to be functional and robust. But to remain functional, infrastructure needs to be resilient in an environment whose stability can no longer be taken for granted. Resilience requires sustainability, particularly environmental sustainability, the ability to continue processes into the future. Sustainable and resilient infrastructure is inherently more equitable, because sustainable energy and processes limit harms compared to unsustainable processes, and resilient infrastructure provides ongoing benefits. Equitable infrastructure means making high-quality and appropriate systems available to more people and communities, and those systems too need to be resilient and sustainable. There are no infrastructural systems, on any scale, that can be functional without also being sustainable, resilient, and equitable. These four characteristics are essentially inextricable from each other.

For communities, countries, and regions that already have full-stack infrastructural networks, an important starting point is decarbonizing these systems. But if that means rebuilding them in their entirety, it's also an opportunity to do things differently. Unlike fossil fuels, renewable energy sources are, by their nature, distributed and decentralized. In the twentieth century, infrastructural facilities moved farther and farther from where they were used as a way of displacing the harms, like the way that coal-powered electricity plants were once in neighborhoods out of cities, but now power plants are mostly outside cities, having gotten much larger in the process. With fewer negative effects on those around them, energy production can potentially return to the neighborhoods they serve. Community microgrids that run on locally available wind or solar or geothermal power and that incorporate storage are resilient as well as sustainable, inherently resistant to disruptions to the electrical grid or to supply chains for fuel.

But full decarbonization requires an entire technological transformation: the replacement of every device, machine, and facility that runs on combustion with an alternative that uses electricity. This has significant implications for materials: Where are all the raw materials going to come from? What about the energy to fabricate them? But also: If we're now thinking about replacing everything anyway, do we really want to do it the same way? Do we replace every gasoline-powered car with an electric car, or do we build out more public transit? Do we replace every natural gas–burning furnace with an electric heater, or can we build out neighborhood geothermal grids? Transforming existing infrastructural systems is like being faced with a Gordian knot but instead of just severing it with a sword, we're figuring out how to untangle it and then how to knit the cord together in a whole new way.

Growing out new systems has a different set of challenges but also a remarkable set of opportunities. Developing countries have the opportunity to leapfrog Global North–style infrastructural networks—the top-down, resource-extracting, CO_2-emitting systems—and instead build out sustainable, local, appropriate alternatives, paralleling the way that many regions skipped over landlines and went straight to mobile phones. Helping resources get into the hands of communities via infrastructural networks is especially important because the places that will bear the brunt of climate disruption are the ones with the fewest resources to respond. And renewable energy is *intrinsically* cheaper than fossil fuels. Both have capital costs to build out as well as running costs, but renewables don't have incremental costs, where every joule of output has to be paid for, because the raw energy input—sunlight, wind, running water, the heat of the earth—doesn't require making a payment to an oil company.

For all their limitations and the inequality baked into them, our infrastructural systems remain an important means to reduce human suffering and to increase agency. It is absolutely possible to create systems that are functional, resilient, sustainable, and equitable, on a global scale. The challenge is figuring out how to get from here to there, technologically and socially.

Black Start

At the very end of our tour of Electric Mountain, our minibus reemerged into the cool gray of a summer day in Wales to skirt the lower reservoir and return to the village at the foot of Mount Snowdon where we began. En route, we passed an unassuming building that blended perfectly into the surrounding landscape of green hills. It was perhaps two stories high, with a steeply pitched roof, faced in the same local slate as the picturesque ruins of Dolbadarn Castle across the lake.

Inside the building is an emergency power system: diesel generators with their fuel supplies, connected to a bank of storage batteries. In the event of a complete systems failure—a blackout across the entire island of Great Britain—it can provide enough electricity to open the valves at Dinorwig, releasing water from the upper reservoir to fall onto the spinning turbines, powering up the station and sending the electricity out to the national grid to bring other systems back online in turn. It's called a "black start."

For humanity, fossil fuels like coal and crude oil were our black start, literally and figuratively. They were conveniently stockpiled underground, easy to access, easy to move around, and easy to use; after all, people have been creating energy by combustion since before we were even *Homo sapiens*. Fossil fuels lifted us out of darkness, literally and metaphorically, giving us the time and resources to create our global, connected, highly cooperative, technological civilization. But above all, fossil fuels created the conditions for humanity to invent the renewable energy technologies that will supplant them. Like the diesel generators at Dinorwig, fossil fuels are a transition phase, not something we want to or even *can* rely on indefinitely. Our civilization is up and running, and now it's time to throw the switch that takes us from the pollution and inefficiency of our black start and powers up a future of sustainable energy and everything that it makes possible.

Nine

AN EMERGING FUTURE
OF INFRASTRUCTURE

Standing in tropical sunlight on a January morning, I had a new appreciation of why it's called a solar *farm*. At higher latitudes, solar panels are often
mechanized in order to track the sun but here, at the SolarArise power plant
just outside of Hyderabad, India, and well south of the Tropic of Cancer, the
person-high panels were just tilted like the gnomons of sundials, each set
facing the sun at the same angle, planted in neat rows, and surrounded by
grassy meadow.

I was being shown around by Prashanth Reddy, the site engineer.
Prashanth's parents lived nearby, and he had starting working at the facility
right after he graduated from a local engineering school. I asked him what
he liked most about his job and he said it was the combination of problem-
solving and spending time outside every day. Much of the work of the plant
seemed to be the kind of quiet routine maintenance familiar to utilities
workers around the world. Here, that included daily washing of the panels,
on a rotating schedule so that any given pane of silicon was cleaned every
month or so. I could see how the just-cleaned row shone beside its dusty
neighbor that was next up to be washed. Inside a control room, two staffers

kept watch for unexpected power drops that would reveal a technical hitch. The entire system is built out of repeating modular units, which makes it easy to narrow down the location and nature of a problem and to swap in new parts. In fact, the whole facility is made up of two identical plants in parallel, built one at a time, first Telangana I and then Telangana II (named after the Indian state we were in). The complex includes shipping-container-sized outbuildings and the transformers that send the output from both carbon-copied halves of the facility onward, via the power cables that I had seen along the road as we drove in, to a substation in the nearby village and then into the electrical grid. The fence around the transformers had signs warning of site-specific hazards. One had a skull and crossbones and the words DANGER 33KV, followed by the same warning in *seven* other languages and scripts—Hindi, Urdu, Telugu, Punjabi, Kannada, Malayalam, and Tamil. The text on the other sign was only in English, but the danger it warned against was quite a bit less abstract and a pictogram of two wriggling snakes conveyed it clearly. I started to watch where I set my feet.

As we walked, I listened as Prashanth explained that the silicon photovoltaic panels don't last forever. They degrade slowly and predictably over a span of years, gradually producing less and less electricity, so their power output is tracked and used to create a schedule to swap in replacements before the efficiency drops below a predetermined threshold. This careful planning can be thrown off by tall grass. Not only does it block the sunlight, lowering the output, but the shadows of the growing, swaying vegetation can turn that steady, even decline of the panels into one that's a lot more random. The solution to this problem was low-tech—the facility contracts with local farmers who come in and cut the grass regularly, keeping the trimmings as fodder for their livestock.

It felt idyllic to be wandering between solar panels in a field out in the country, everything around me peaceful and just silently functioning. When I left New Delhi that winter morning, it had been dark, freezing, and smoke-saturated, but here the day was warm enough that I had shed first my coat, and then my scarf and sweater, and was now down to a black T-shirt, absorbing solar radiation and turning it directly into warmth on my skin. I remembered the old joke and I shared it with Prashanth: What happens when there's a fuel spill at a solar power plant? It's a sunny day.

It didn't feel like the future. It felt like the present.

The Narrowest Future

Postapocalyptic futures are a staple of science fiction, and almost by definition those futures don't have well-functioning infrastructural systems. In popular books, film, and TV, there are hardly any positive, utopian visions of the future where everyone's needs are met by largely invisible systems. If there is a society like this, the action usually takes place at the periphery, like Starfleet in *Star Trek*, or it's only for a small elite, resulting in conflict between the utopian haves and the dystopian masses of have-nots, as in *The Hunger Games*.

I often joke that the only science-fictional future that I personally want to live in is the one where everyone lives a life that is long, cared for, and well provisioned, so that the worst thing that could happen to you before your peaceful death is heartbreak. Those stories don't exactly lend themselves to exciting, big-budget, CGI-loaded movies. One consequence of their absence from fictional narratives is that most of us don't have a lot of practice thinking about the shared futures that we *do* want to live in. I'm comfortably middle-aged and in good health; barring death by misadventure, that means I can reasonably expect to see the middle of the twenty-first century. But any estimate of my life expectancy presupposes that I'll have continued access to the kinds of infrastructure I do today, in an urban area of the American Northeast, like clean water and adequate sanitation, adequate heat and cooling and reliable energy, to say nothing about the health care and other systems they enable, or the supply chains and food systems we depend on. Access to these systems contributes to the variation in life expectancy by where you live, and it's also why excess death rates rise sharply when these systems fail, as during long-term power outages.

What if all I cared about was my *personal* access? What would it take to keep the infrastructural systems that I rely on functioning throughout the next four or so decades of my own lifetime, so that my future continues to resemble my privileged present? A fair bit of work, it turns out. These systems were mostly built to be robust and redundant, but they'll also need to be resilient and responsive so that they can continue to function as the effects of climate change become more and more apparent.

By 2050, the mid-twentieth-century building boom in the U.S., Canada, and other places will be a century in the rearview mirror, and entropy will have been taking its toll that whole time. So the obvious place to start is to buttress existing infrastructural systems, starting with finally doing all the maintenance that's needed to bring them back into good repair. That would mean systematically and swiftly replacing leaky century-old water pipes, bringing aging bridges back up to structural snuff, going through the daunting and multivarious repair backlogs of transit and transportation systems of all sorts—everything that's been pulling down the grades on ASCE's national infrastructure report card. And not just locally, because my life isn't restricted just to New England and neither are these systems.

But we already know that routine maintenance alone won't be enough to get us to the middle of the century. Keeping these systems functional means starting to deal with the threats that are already here, including greater variability in the weather. It means winterizing the electrical grid in places where the weather is getting unexpectedly cold, like Texas, and it means upgrading maintenance and tree-trimming schedules and installing equipment to reduce the risk of fire from energized power lines in places that are becoming increasingly hot, not just California but the entire West Coast, at least as far north as Vancouver. It means planning for heavy rain and building strategies for flood mitigation from New York City subways to dams across the rural Midwest. It also likely means *unbuilding* structures, whether that's relocating buildings away from shorelines facing sea-level rise and higher storm surges, or taking up the concrete around urban rivers and replacing it with bioswales, vegetated channels that absorb stormwater, preventing floods while removing pollutants. It means upgrading water supplies to manage emerging classes of contaminants, like PFAS, and it means hardening the electrical grid continent-wide so it can weather the next Carrington Event–level solar storm.

All of this is just the absolute minimum to maintain the status quo for another few decades, and much of this work is already happening. But even with this narrow perspective, there are obvious concerns. First, no matter how short our time horizon, it doesn't make a lot of sense to just blindly invest in rebuilding systems without considering whether they're the systems we want to have, potentially for decades to come. If you're digging up and replacing leaky water mains, it's a good time to think about separating out stormwater and sanitary sewage so that wastewater treatment plants aren't overwhelmed every time it rains. Repairing roads and other transportation routes? It's an opportunity to address congestion, parking and land use, and traffic safety, which means considering alternatives to private vehicles. As always, maintenance shades into renewal and rebuilding and replacement. But also, there's no way to maintain a status quo when the world is changing around it. Routine challenges, like bad weather, are starting to become anything but routine, as climate instability leads to increasingly extreme weather events. But it's not the weather events that are the natural disasters. A blizzard in Montreal is par for the course; the same blizzard in Atlanta is a different story. What makes a natural disaster is a severe adverse impact on a community and its infrastructure, and the more vulnerable they are— both the people and the systems—the more severe the impact. But this also means that this impact can be lessened by creating resilient infrastructure, including the social systems around it. Planning and building for the next half century of climate change means planning to mitigate natural disasters.

Reliability and Risk

The ferry is my favorite way across the San Francisco Bay, but it's also crossed by a number of bridges and the Transbay Tunnel that carries the BART trains. In the aftermath of the 1989 Loma Prieta earthquake, which was strong enough that the Bay Bridge was seriously damaged, that redundancy in the regional transportation network meant that ferries could bring people in and out of San Francisco during repairs to the bridge. Redundant networks with multiple paths are intrinsically resilient. Conversely, where the network redundancy is low, the system as a whole becomes brittle. This is what's different about snow days on the eastern and western ends of Interstate 90. A traffic accident on the highway in Boston, where it's embedded in a web of surface roads, is a minor inconvenience. In Snoqualmie Pass, through the mountains just east of Seattle, shutting down a lane or more of the highway means a major disruption to travel in and out of the city. This is also one of the fundamental design principles of the Internet, with its decentralized structure and network protocols that allow for routing around damaged nodes. A similar network redundancy in the electrical grid is why power could be routed around the damaged Metcalfe substation after it was attacked, avoiding major outages.

When there's a single or just a few points of failure, they need to be reliable, not just in the face of routine challenges but also under extraordinary conditions. That high reliability can be achieved by a combination of making facilities robust and incorporating redundancy, like the triplicate systems of my local water treatment plant. But in a world of climate instability, we're already beginning to see that might not be enough. Faced with conditions that are challenging and unpredictable, systems need to be resilient. A brittle system is one that "falls over," failing all at once. Resilient systems can remain largely functional even if some individual elements fail, or else can recover their functionality quickly after failure.

Resilience is strengthened by redundancy and also by diversity, incorporating different *kinds* of elements. It's why scheduled service for a ferry fleet on the San Francisco Bay can be an aspect of disaster resilience in addition to being a daily commute. In the case of an earthquake severe enough to damage bridges and tunnels, an alternate transportation mode across the bay would be critical, especially in the crucial first hours and days of disaster response. Resilience is also strengthened by decentralization, with smaller, distributed facilities rather than just a few large ones, however robust. Here's where renewable energy has the potential to really shine, because the energy inputs (water, wind, sun, geothermal) are distributed by their nature, unlike fossil fuels. Local electricity generation can be coupled with local energy storage, like batteries, configured as a community-level microgrid, and connected to the main electrical grid. Whether there's a severe and widespread event like a hurricane or a small-scale issue like a substation failure, the microgrid could continue to provide power even if the connection to the larger grid goes down, and it's not dependent on fuel deliveries or other supply chains. And unlike diesel generators, having a community-scale microgrid means that everyone is already connected to their neighbors and can easily share emergency power. Mutual aid for disaster response is built into the system itself. In cities, where residents mostly use electricity brought in from generating sources elsewhere, one starting point is "resilience hubs" that can provide power just in emergencies. Electricity cooperatives in rural areas, which already have local distribution networks in place, are leading the transition to renewably powered, resilient microgrids.

What's common to all of these approaches—robustness, redundancy, and resilience, especially through diversity and decentralization—is that they are not efficient. Making systems resilient is fundamentally at odds with optimization, because optimizing a system means taking out any slack. A truly optimized, and thus efficient, system is only possible with near-perfect knowledge about the system, together with the ability to observe and implement a response. For a system to be reliable, on the other hand, there have to be some unused resources to draw on when the unexpected happens, which, well, happens predictably. This is what safety factors in the design of bridges and other structures are for: they are "overbuilt" (in careful, specific ways) to compensate for uncertainties that can't be eliminated entirely. Electrical grids have significantly more power on tap than the predicted demand. In fact, having a reserve is often required by regulators to help make sure that the grid is reliable. Winterizing an electrical grid is about expanding the temperature range in which it functions well, but on most days, that winterization isn't needed. The same is true of retrofitting systems, like rail and power lines, so that they continue to function well at the hottest temperatures. Redundancy is a form of slack; passenger jets have multiple engines so that they can still fly safely if one fails. Sometimes the slack is in the form of a buffer, a supply of whatever is being moved around. A battery is a buffer, and so is a reservoir that holds water until it's needed for irrigation or power generation.

Sometimes that slack is provided by people. An overbooked flight is one that's been overoptimized, and so when a gate agent asks for volunteers to give up their seat in favor of a later one, it's a way of finding some human slack in the system. The widespread use of real-time traffic mapping services has become a source of slack—of flexibility or resilience—for roads, as drivers use that information to disperse out along different routes. But most of the time, when infrastructural systems fail, humans are not so much providing slack as they are dealing with the consequences and paying the price. A blackout might shut down a factory, but a five-day blackout during a winter freeze in Texas shuts down *everything*—the bodies and lives of the residents serve as the buffers and the reservoirs and the source of resilience. When a municipal water supply is temporarily unsafe to drink, that means that someone is spending part of their day boiling and cooling drinking water or getting bottled water to drink and for hygiene. When it's unsafe to drink indefinitely, this work goes on indefinitely.

Optimizing a system means having some metric for what it means to run well, as well as a predictable context—what it's being optimized for, and under what conditions. A plane that's completely full of passengers is optimized for cost—in principle, it should maximize revenue generation for the airline, or minimize ticket price for the customer, or both. But if a thunderstorm grounds the plane, that means that this optimized system has now become a plane full of people who don't have a way to get home, especially if all the later flights are also similarly "optimized," with few empty seats. Because infrastructural systems are mostly about nonmonetary benefits to people—their time, agency, comfort, and health—optimizing them for price, much less revenue and profit, misses these important aspects. Systems can only be optimized for what they measure—the cost of providing electricity can be accounted for, but missing from that accounting is what it means to spend a day huddled under blankets because the heating is out during a winter freeze—and only under predictable conditions. Climate instability means more uncertainty in the larger context of infrastructural systems. As the environment in which they're embedded becomes more variable, that means de-optimizing these systems. It means investing in making them reliable under a wider and less predictable range of conditions and putting in the safeguards necessary to prevent outlier events from becoming natural disasters, where the highest human costs are paid by the most vulnerable. If the twentieth century was about building infrastructural systems that could manage or even just displace risks, the twenty-first century is going to be about building out systems—human as well as technological—that are resilient enough to sustain our communities during conditions of ongoing uncertainty.

Responsiveness and Adaptability

I was born after most of the systems that I talk about in this book were first built out in the places where I've spent most of my life: municipal water provision and wastewater treatment, a continent-wide electricity grid, the telephone system, public transit systems, utility gas distribution in cities, the highway system, civil aviation, and containerization. The one big change has been telecommunications, with the rise of the Internet and of mobile telephony; even then, much of the domestic-scale physical networks, like phone lines and telephone cables, were installed in residences long before they became repurposed to provide Internet access. Many of the systems that I use today were built out using a top-down, technocratic mindset that imposed them on communities that had little say in where they would go, how they would be used, and by whom. It's highways through African American neighborhoods and reservoirs that drowned ancestral hunting grounds. It's also huge centralized systems like the Deer Island treatment plant or the Pickering Nuclear Generating Station.

This approach arose from a certain epistemological stance: the planners and engineers who were designing and building out infrastructural systems believed that they knew what to do and how to do it, and that it was the right thing to be doing. Centralization and robustness—designing for scale and longevity—follow from this. Even at the time they were built, those beliefs were challenged—that's why there's a subway line and a linear park heading southeast out of Boston instead of an expressway—but now, half a century or more later, we have a very different perspective on the decisions that were made and the tradeoffs and biases that were built in. What's more, in a world with a rapidly changing environmental and technical context—climate instability and the rapid development of new technologies, particularly in energy generation and storage—it doesn't make sense to make big, irreversible investments. Since we don't yet know exactly what we're going to need or how it'll work, what makes more sense is smaller-scale, distributed, reversible systems. It's a counterintuitive way to think about systems that are characterized by continuity, but in order to keep functioning in a changing world, these systems will need to change with time—to be *responsive*. More responsive infrastructure, systems that incorporate diversity and decentralization, can also be more resilient. And a more provisional, adaptive approach to building out infrastructure, one that involves exploration, experimentation, and prototyping, provides pathways to getting community input, seeing how new facilities and systems work, and then modifying them into something different.

Some of this responsiveness was in evidence in cities during the first year of the coronavirus pandemic. I had bought a bike and had started riding around town not long before the public health crisis began, prompted by the way my neighborhood was steadily getting more bike-friendly, including the installation of bike racks and repair stands. When I was biking home from the library where I often worked, my commute would take me across a bridge that goes from a busy street in Boston onto one in Cambridge that's nearly as busy, but instead of dodging cars, I'd find myself in a protected bike lane, complete with bright green road markings. Crossing over the river on my bike always feels like being in the kind of fantasy novel where the heroes face enemy resistance on their journey but magic-eased passage across their own kingdom.

But what was really striking was how much new cycling infrastructure was put in place during the pandemic and how fast it went up, as the built environment rapidly became the "build it as you need it" environment. On both sides of the Charles River, there were new, temporary bike lanes, often initially marked off only with large orange and black safety barrels and temporary signage. Busy commercial streets saw traffic lanes repurposed, not only for bike lanes but as "slow streets," as pedestrian thoroughfares, or to create extra-wide patios for outdoor seating. Over the course of the pandemic and the turning seasons, those patios were first augmented with heaters, and then the tables and chairs were put into storage as winter arrived in earnest before being brought out again in the spring. The coronavirus pandemic provided both a need for and an opportunity to experiment with different ways of using city streets, allowing the community to "try out" temporary responses before they became permanent changes to the local landscape. Urban transportation activists have long advocated similarly experimental and responsive strategies with bus routing, taking advantage of the inherent flexibility of buses on roads.

Temporarily modifying the way that streets are used is a pretty minor example of how an infrastructural system can be locally responsive to changing external circumstances. We start to see the real power of systems that are both resilient and responsive, that are reliable but flexible, robust but not brittle, as we start thinking about longer timescales and get further and further away from the present and our current way of doing things and into a world with more unknowns.

Sustainability

I don't have children of my own, but I do have my niece and three nephews, and I get to be an honorary auntie to the children of many of my friends. If the rate of infant mortality in your community becomes low enough that you start expecting that newborns will make it to adulthood, the future seems a lot closer than it ever did before. That next generation is on track to see the twenty-*second* century if they live as long as their parents and grandparents, but their life expectancy is just as dependent on access to collective infrastructural systems as mine is. If we want those systems to function throughout their lifetimes and into the lifetimes of our grandchildren, we are easily thinking about a century or longer. To be fair, there are plenty of century-old infrastructural systems still in use today, from municipal water supplies to telephone lines to railway stations to pipelines. As the descendants of their creators, we still benefit from their investment, of care and expertise as much as money. But those systems were mostly built out under two closely related assumptions. The first is that the resources they were tapping—most importantly, energy in the form of fossil fuels, but also water, and other materials—were limitless. And the second is that the planetary ecosystem is an infinitely receptive sink for waste and pollution, capable of absorbing any amount without being altered.

We can no longer even pretend to ourselves that these assumptions are grounded in reality.

Burning fossil fuels for the energy to power infrastructural systems has planetary-scale impacts, and the total amount of carbon dioxide and other greenhouse gases in the atmosphere is increasing faster than it ever has in human history. Other waste from industrial-scale processes, including microplastics and "forever chemicals" in water, is also globally distributed, found far from where it's produced or used. Fossil fuels also have localized impacts, from mining to fracking to pipelines to refineries and chemical processing plants. The environmental depredation at these sites of extraction and processing, and where the end products are incinerated or land-filled, have long been considered "out of sight, out of mind," as supply chains and transportation networks have made it possible for them to be farther and farther away from the point of use. The rise of global mobility and telecommunications in the last fifty years has gone a long way toward erasing the idea that there is an "elsewhere" or "others." These sacrificial zones are no longer entirely invisible and the voices of the communities that are affected are finally being heard widely, in a way they've never been before. There is still a tremendous ongoing challenge in making the connection between behaviors and outcomes, much less establishing alternatives, like "right to repair" regulations that make it possible to fix electronic devices and keep them out of the e-waste stream. But day-to-day decision-making aside, once we begin to think about what these systems might look like a century from now, we can no longer plead ignorance or delude ourselves about the negative impacts at either the local or the planetary level. After a century or more of drawing on resources as if they were inexhaustible, we now understand that our infrastructural systems need to operate very differently. They need to be sustainable.

I hear the word "sustainable" so often and used in so many different ways that I can easily lose sight of what the word actually means. Sustainability is fundamentally about time, which means it's about verbs, not nouns.

For something to be sustainable, the processes or practices involved can continue to work, in the same way, into the future. How long into the future? Well, it depends on the process and its context: maybe it means a few months or a few years, or a few decades or, as here, a century or more. So how do we know that something is sustainable? The short answer to this is that we don't. But the shorter the timeline in front of us, the more confidence we can have about whether a process is sustainable for the duration. I'm a casual runner, but I am reasonably sure I could go out the door and run three miles. Six miles would be more of a stretch. I've run half-marathons and even a marathon before, but I'm not at all confident that I can run one now, certainly not without a lot more training. And maintaining even a slow jog for a hundred-mile ultrathon feels pretty out of reach. There's no way of knowing something is sustainable on an appropriately long timescale, because there's no place to stand and look back. But while we can't know for sure that something is sustainable, we often can be quite sure when it's *not*.

If one element of sustainability is time, the other element is the process itself—what exactly is it that we are continuing to do the same way into the future? Is it the right process to sustain? I don't think I could run for a hundred miles, but I'm pretty sure I could walk a hundred miles if I had to. And I've never done a century ride, but if I needed to cover a hundred miles under my own power, I would get on my bike. As long as I was cycling, I could probably go farther with less energy than if I were on foot, but if I couldn't cycle, I could still walk, and maybe even push my bike alongside, gracefully transitioning from one process to a different one. I am most confident about my ability to drive a hundred miles—it's less than two hours at highway speeds—but that absolutely presupposes and embeds so much more than being on foot, starting with access to a highway, to my car, and to the fuel needed to power it, and it opens up a whole other web of externalities like air pollution and road safety. Deciding whether something is sustainable is deciding what consequences we're willing to accept, when, and for whom.

Before we address the more complex question of what it might take for our infrastructural systems to be sustainable for a century or longer, we can start with the much simpler question: What systems are *not* sustainable on this timescale? That means considering whether the process itself can continue for the timeline in question, and also what happens if we keep doing things in the same way.

Beyond Unsustainable Systems

Any infrastructural system that involves extraction of nonrenewable resources or dispersal of pollution is, by definition, not sustainable. The exact timelines might be up for debate, but after a century or so of industrialization, we possess a deep industrial, economic, societal, and scientific understanding that even seemingly inexhaustible resources have limits, and seemingly infinite pools for dilution can be contaminated.

In the familiar water cycle, it falls to the ground in the form of rain or snow, and may be used by humans for drinking or irrigation or industrial purposes, before eventually evaporating, condensing into clouds, and falling to the earth again. But not all the water we use is part of this system, at least not on human timescales. Researchers from the United States Geological Survey established in the 1960s that the Ogallala Aquifer, the subterranean source of the water used to irrigate those circular fields across the midwestern U.S., recharges so slowly that it's functionally nonrenewable. The drop in the water level has since become unignorable. It varies from place to place, but in some parts of Kansas, it's been as much as 150 feet, deep enough that farmers had little choice but to abandon wells. While there's no single answer to the question of when it's going to run out, it's likely to be far sooner than our working horizon of a century.

Unlike water from the Ogallala Aquifer, the unsustainable aspect of fossil fuels lies not only on the extraction side but on the waste products side. Combustion transforms coal, oil, and natural gas from hydrocarbon chains buried deep in the ground into carbon dioxide and other greenhouse gases, and it's the effects of that increasing concentration that are becoming untenable. If terraforming is taking an uninhabitable planet like Mars and changing the atmosphere to make an ecosystem capable of supporting life, we are instead taking our perfectly habitable planet and veneriforming it, transforming our terrestrial home into our other planetary next-door neighbor, suffocatingly hot Venus. In the near term, it might be enough to focus on making existing infrastructural systems more robust, even if we don't change how they are designed and built. But on longer timescales, certainly out to a century or longer, the consequences of uncontrolled greenhouse gas emissions—allowing them to continue to accumulate in the atmosphere— are going to affect not just our infrastructural systems but virtually everything else about our planetary biosphere, including every living being. The continued use of fossil fuels is unsustainable because these are consequences that we should not be willing to accept.

If weather systems become increasingly chaotic and extreme, and temperature increases mean that places where humans currently live become increasingly inimical to life, then what kind of infrastructural birthright are we leaving for our descendants? If we think about infrastructure as increasing our agency and the number of choices available to us, then abdicating our responsibility to leave them a world where they can live lives of peace and stability, with their basic needs cared for, is a failure of intergenerational empathy on a civilizational scale. Our current practices are unsustainable because they *reduce* the choices and agency available to coming generations.

If we want any chance of having functioning infrastructural systems a century from now, we'll need to try to stabilize—or at least keep from further destabilizing—our ecosystems by limiting greenhouse gas emissions, which means decarbonizing our energy sources. This is why we mostly (but not always) use the terms "sustainable energy" and "renewable energy" interchangeably: powering our infrastructural systems with renewable energy at least has the potential to be sustainable for a century or longer. But reaching that goal means successfully transitioning them away from the combustion of fossil fuels. Is it even possible—is there enough available renewable energy to power these vast systems? What technologies are needed? And there's an even bigger question: How do we transform infrastructural systems at scale, collectively, while we are depending on them? It's an enormous challenge, but it's also an unprecedented opportunity. Not only is it a pathway to a sustainable future, but that future is one of energy abundance, not scarcity.

An Energy Accounting

What would it take for every human on the planet to use about the same amount of energy as I do, in my warm, well-lit home, with fuel for cooking on tap, and access to mobility options including a personal vehicle, mass transit, and air travel? We can ballpark it by taking the per capita energy footprint of Canadians and multiplying it by the eight billion people in the world, which gives us about 2.4×10^{21} joules per year, roughly ten times the current global energy consumption. That's the total amount of energy that would be needed each year for every person on the planet to have access to the same generous amount of energy as the average Canadian (I used Canada, not the U.S., because it's similar but both colder and larger). While worldwide energy consumption is increasing over time, the amount of energy used in wealthy, industrialized countries like Canada has been pretty flat for the past few decades, which suggests that energy usage plateaus. The global growth in energy consumption over time is mostly driven by economic development—more people getting access to more resources, and hence to more energy.

Estimating the total energy usage of humanity, with all the different and changing ways that it's used by people all over the world, is an ongoing challenge. Estimating the amount of solar energy landing on Earth, on the other hand, is a relatively straightforward question of experimental physics and geometry. In its orbit, eight light-minutes away, Earth intercepts only a tiny fraction of all of the energy coming off the Sun but it's enough to bathe our planet in radiant light and warmth to the tune of about 5.5×10^{24} joules each year. That means that our estimate of the civilizational energy usage to meet the needs of every human on Earth is about 0.04 percent of the incident solar radiation (which also powers the movement of wind and water on the planet). Four parts in ten thousand. The tiniest fraction of the sun's energy could power all the infrastructural systems we use today, and probably a good long while into the future. That means it's at least *feasible* that our energy usage, as a species, could be sustainable, in the biggest possible picture and the longest possible time frame. This long-range view shows us that our fragile blue-green marble of a planet, suspended in space, is absolutely awash in energy—enough to provide all humans with the power to live comfortable lives of agency and connection.

We're accustomed to thinking about reducing carbon emissions as going hand in hand with reducing energy consumption. But knowing that there is enough renewable energy to meet the needs of every human on Earth is a big deal, because it changes the terms of the problem. Rather than focusing on *reducing* our energy usage—a scarcity mindset—we instead can focus on *increasing* the amount of energy that comes from renewable sources. And we have a suite of emerging technologies that allow us to get energy without setting anything on fire, which means that for the first time in human history, it's possible to entirely decouple energy usage from carbon emissions. If we put these together, we can realize that stabilizing the global climate doesn't have to mean rationing or limiting energy usage, and it doesn't have to mean sacrifice. Transitioning rapidly and completely into renewable, decarbonized sources of power is how humanity can access unprecedented energy abundance.

It's possible to power our entire technological society on renewable energy. The next step is to figure out how.

The Energy Transition for Infrastructure
(and Everything Else)

Switching to renewable sources requires harvesting this energy, which is typically diffuse or intermittent, and transforming it into a form where it's reliably available to do useful work on demand. That means technologies for generation and also ways to concentrate, store, and distribute that power. That's the upstream part—transforming the supply and distribution of energy.

The downstream part is usage. Everything in our lives that currently uses fossil fuels directly, like heating, cooking, and transportation, will instead need to run on electricity. When natural gas distribution first arrived in New York in the 1950s, technicians went from kitchen to kitchen to manually adjust every single stovetop burner and oven element in the city because natural gas burns hotter than the coal-derived town gas that it was replacing. Figuring out how to effectively use renewable energy sources means not only developing the technologies and building out the new systems but, in much of the world, it means managing the conversion. All the vehicles, appliances, and other equipment that treat fossil fuels as a utility will need to be replaced or retrofitted.

This transition would require the near-total transformation of three infrastructural networks. Two of them rely on fossil fuels for direct combustion at the point of use. One is the utility gas piped to buildings (here in the northeast U.S., it's a primary energy source for heating as well as for hotwater heaters and cooking) and the second is transportation, since vehicles are almost entirely powered by internal combustion. Most of this energy would need to come from the electrical grid, which is the third, and overarching, system that would need to be transformed. Decarbonizing energy infrastructure means retiring thermal generation that relies on burning fossil fuels (coal, oil, or natural gas, as well as temporary sources of electricity like diesel generators) and replacing it with electricity from renewable sources. It also means growing this electricity supply enough to replace all the energy that's currently coming from fossil fuels at the point of use.

Diffuse, Diverse, and Distributed

An important difference—maybe *the* important difference—between fossil fuels and renewable energy sources is that the former are concentrated and the latter are diffuse. This is true in both physical and geographical terms. Fossil fuels are energy-dense, which is why we've built transportation systems around them. It takes about seventy-five seconds to fill up the tank of my car and the gasoline weighs less than a hundred pounds. Even though only about 30 percent of the energy from combustion gets turned into forward movement, it's still enough for nearly six hours of travel at highway speeds. Fossil fuels are also concentrated in certain places in the planet—the coal mines of Appalachia, Newcastle, and Inner Mongolia; the oil fields of Texas, Saudi Arabia, and western Siberia; the oil sands of Alberta; the North Sea reserves off the coasts of Norway and Scotland. Renewable energy sources, like solar, wind, hydropower, and geothermal energy, have a much broader geographical distribution, but they're also diffuse. In order to be used effectively, the energy needs to be captured and concentrated. That builds in a strong bias toward using the renewable sources that are most locally abundant or accessible—solar power in tropical India, geothermal for heating and electricity generation in seismically active Iceland, hydropower in Alaska, wind turbines in Oklahoma.

While the technologies continue to advance, particularly around increasing efficiency, thermal generation of electricity using coal or natural gas is a mature technology, and plants can be built pretty much anywhere in the world that can get a steady supply of fuel. In contrast, diversity of renewable energy sources rewards a diversity of approaches to harnessing it, which means the technologies tend to be specific to where they're located. They're also in different degrees of development, although they're generally less mature than the fossil fuel technologies that have a century or more of industrial research and practice behind them. In February 2020, I finally got the chance to take a tour of the wind farm at San Gorgonio Pass, outside Palm Springs, having driven past the rows upon rows of gently spinning white turbines a number of times. Exactly like Electric Mountain, the tour began in a trailer where a guide who'd had a long association with the site gave an overview to our group of interested tourists, mostly retirees, before we all piled into a minibus to get a closer look at the turbines. We also had a chance to look around what I could only think of as a wind turbine graveyard, where failed examples of designs from decades past had been laid to rest. That impression was heightened by the informational markers, which were the tips of turbine blades embedded upright in the desert ground, so that the squared-off base and arched top echoed the classic headstone shape. Each was labeled with details of the design, the dates the turbine was operational, and the mode of failure (the cause of death was mostly that they ripped themselves apart while in use). Most of the active turbines were all very similar, because it's a largely mature technology—they're based on what's known as the "Danish design," with its slender, gracefully curved blades, a self-centering nacelle, and a gearing system that allows it to work efficiently and safely at a range of wind speeds. The turbine was developed and refined with the support of public funding from the government of Denmark and policy incentives that resulted in the widespread adoption of wind power, the majority via wind energy cooperatives. Solar panels are similarly mature, due in large part to decades of government funding and the global policy support which shepherded higher-efficiency panels out of the lab and into large-scale commercial production, while reducing the cost of installations. Geothermal power systems are already in use in seismic hot spots, especially along the Ring of Fire encircling the Pacific Ocean. Technologies to harness the heat of the earth from a wider range of locations are under development. Renewable energy projects leverage existing expertise and equipment, particularly from the fossil fuel industry, borrowing drilling techniques for geothermal power or repurposing the highly specialized ships that are used to transport and set up offshore oil rigs to instead build out offshore wind turbines. All of these technologies are continuing to develop and diversify, a Cambrian explosion of sustainable power.

A wind farm or a solar plant typically has a much bigger physical footprint on the ground than a conventional thermal generating plant that produces the same amount of electricity. Large-scale renewable energy installations are often built up of modular parts, spread across the landscape, like the San Gorgonio Pass wind farm, with its row after row of turbines. At the solar farm in Telangana, not only were the individual solar panels modular and swappable, but the whole plant was built out in two halves. That modularity also lends itself to different scales and to distributed generation, in the form of many smaller installations rather than fewer, larger ones. Thermodynamics considerations mean that thermal generators generally get more efficient as they get bigger, so they're built as large power plants, not by daisy-chaining together smaller turbines. Decentralized and distributed generation of power from fossil fuels looks like a neighborhood filled with loud, polluting diesel generators. (It also looks like one filled with private gasoline-burning cars, of course.) Distributed generation of renewable energy, on the other hand, might look like a neighborhood with solar panels on every roof, or with heat pumps, or with small wind turbines.

Transitioning to renewables also means learning to use the energy that's present in the local environment. I visited an experimental, small-scale, run-of-river hydroelectric power turbine in Pawtucket, Rhode Island, just downstream of the Slater Mill, a water-powered cotton mill that dates back to 1793 and is considered the first factory in the U.S. (and the site of the first factory strike in the U.S., led by young women workers in 1824). "Run-of-river" means that there is no reservoir—instead, a fraction of the flow of water in the river is diverted to the turbine, used to produce electricity, and then returned. In order to limit the environmental and ecological impact, a minimum water level in the river is maintained, as at Niagara Falls. The amount of power generated then depends entirely on the flow, more in the spring when the water is flowing fast with melting snow and less from the sleepy river of late summer and early autumn. In contrast to the massive reservoirs of traditional hydroelectricity, the ecological disruption is minimal, which means that many more sites are viable. Small-footprint, run-of-river hydro is already a significant fraction of electricity production in mountainous British Columbia.

This list of technologies for harnessing energy in the environment is by no means exhaustive. They're all sufficiently developed that we have reason to believe that decarbonizing our infrastructural systems is technologically possible, but not so far along that there aren't still new ideas and new possibilities emerging. By its very nature, renewable energy is likely to be much more diverse, using a range of energy sources in a variety of ways according to what's locally abundant or accessible.

From Supply to Storage

All things considered, demand for electricity is surprisingly predictable—the way it changes with the time of year, with that diurnal "duck curve," and with the weather. Decades of experience in how people use power, service areas that are large enough that individual differences are absorbed into an aggregate whole, and reliable weather forecasts make it possible to accurately predict the minute-by-minute demand for electricity.

The reliability of the electrical grid comes from just-in-time energy production, with careful oversight in grid operations centers to ensure that generation is closely matched to consumption. In large hydroelectric power stations, the water impounded behind the dam serves as a reserve of energy, powering a set of turbines lined up in parallel, with individual turbines spun up and down as needed during the day. Nuclear plants, on the other hand, produce large amounts of power but the amount can't be changed quickly, so they typically supply the stable base load. It's why they pair beautifully with fast-response pumped-storage hydroelectricity stations, like Dinorwig. Powering the pumps that reset the reservoirs with the "excess" energy from nuclear power during the overnight lull in demand hits the trifecta: it's efficient, inexpensive, and low-emissions. By and large, though, the ability to match electricity supply to demand has depended on the use of fossil fuels, because they can make energy generation continuously scalable, in much the same way that putting your foot on the gas pedal makes a car go faster while easing up on it means the car slows down. Thermal generating stations are vastly more complicated, of course, but the basic principle still applies: control of the timing and amount of fuel input means full control of the electricity output. It's part of why peaker plants, which handle the intervals of highest electricity demand, typically run on natural gas or petroleum.

But with renewable sources, the available energy can vary considerably over time. Some of that is predictable: hydropower, wind, and solar power all vary seasonally in predictable ways, and sunrise brings every night to an end, at a time that is calculable and inviolate. Superimposed over these patterns is the daily unpredictability of the weather, whether it's clouds over solar panels or a windless day that becalms all the local turbines. The moment-by-moment supply of renewable energy, therefore, can be intrinsically uncontrolled and somewhat unpredictable. Since insufficient supply to meet demand leads to brownouts or even blackouts, this feeds into concerns (or fearmongering) about the "unreliability" of renewables. A supply that's much greater than the demand can also be a problem—at worst, unplanned-for electricity means power surges in the grid.

Switching to renewable energy, then, means figuring out a way to reliably meet demand without having the fine-grained control of supply that fossil fuels provide. The first step is already part of conventional generation: aggregating the electricity supply into a shared grid. This lets us average out unpredictable variation across different places—it might be cloudy in San Francisco but sunny in Los Angeles, so we can share solar power between them. It also allows switching between different modes of energy generation.

The Mojave Desert is home to not just the wind farm in the San Gorgonio Pass but also a number of large-scale solar projects, which means that cloudy days might be offset by windy ones, and sunshine helps to maintain the power supply on calm days. But effectively using renewable sources also means capturing the electricity at the moment it's produced and keeping it available for use when demand is high or because weather events like *dunkelflaute*, "dark doldrums," periods of cloud cover with little wind, mean that production is low. This requires energy storage, whether it's for hours, days, or even weeks.

Pumped-storage hydroelectricity plants are the granddaddies of grid-scale energy storage, a widespread, mature technology for storing large amounts of energy indefinitely. At the other end of the size scale for electricity storage are the ubiquitous lithium-ion batteries that power portable electronics, made larger to store proportionally more electricity. Rechargeable batteries can hold their charge for long periods of time and they're energy-dense. I keep an emergency battery pack in my car that packs enough of a wallop to jump-start the engine but is still small enough to hold in one hand. Rechargeable batteries have significant downsides, though. First, they're expensive. More than cost, however, is the issue that battery production for the grid is in competition for materials and fabrication capacity with consumer products. This includes electric cars, however, and vehicle batteries might end up being grid-scale storage in a roundabout way: they can act as a distributed storage grid for renewables, including serving as an emergency power source. But the biggest concerns, by far, are the geopolitical issues and environmental costs associated with mining and refining the raw materials for batteries—not just lithium but also cobalt and even copper. All of these concerns will get a lot worse if production scales up without reconsidering how to handle extraction, disposal, and recycling, or developing alternative ways of storing energy.

Just as renewable generation can match the specific conditions where the energy is being generated, storage can too, including co-locating it alongside. I walked through fields of solar panels at the Telangana plant, and the San Gorgonio Pass wind farm stretches for miles alongside Interstate 10. Energy storage in places like these doesn't require the compact portability of batteries—quite the opposite. One possibility is gravity storage, where a crane equipped with an electric motor and a cable stacks up concrete blocks and is then run "backward"—as a generator—when the blocks are picked up from the stack and lowered back down to the ground. It's exactly the same principle as pumped-storage hydro, except that the potential energy is stored in concrete blocks instead of water, so it can be built out almost anywhere and with common materials. New technologies under development include chemical batteries that are optimized for reversible storage and longevity instead of energy density. Researchers at the Massachusetts Institute of Technology have been developing a number of different battery systems, including one that uses plain old iron oxide and a reversible rusting process, storing energy in chemical bonds and then running the reaction in reverse to produce electricity when needed, months or even years later. Like renewable energy generation, some grid-scale storage technologies are already feasible but there's also a whole a host of emerging technologies, different ways of filling different niches.

Decarbonizing Heating and Cooling

In North America, direct combustion in the home is used for three main purposes: cooking, home heating, and hot-water heating. The fuel is mostly utility gas, but heating oil is common for homes in New England and elsewhere that aren't on a municipal gas grid. Burning fossil fuels to produce heat to generate electricity, delivering it, and then turning it back into heat is less efficient and therefore usually more expensive than in-home combustion. A decarbonized grid drastically alters the economics of using electricity to generate heat. Not only does it sidestep these inefficiencies, it can mean *no longer paying for fuel at all*. This model is already in effect for places like Quebec. Despite the bitterly cold winters, electric heating is common because the public power utility generates cheap, abundant hydro-electricity. It can also be an economically compelling argument for energy-hungry commercial users. Lava Glass, a glassblowing studio in Aotearoa New Zealand, has gone carbon-neutral, including replacing gas-fired furnaces with electric ones powered by their local utility, produced from hydro-electric and geothermal plants. Since the energy costs of the huge quantities of heat produced by the furnaces and the annealing ovens are a major expense for any hot shop, the switch from gas to local renewable electricity means not only lowering their operating costs but also stabilizing the costs and the supply. It's easy to lose sight of this in the complexities of policy and utility pricing, but whether it's at the scale of solar panels on the roof of a single-family home or the entire electrical grid, making an upfront investment in renewable generation is intrinsically related to reduced day-to-day energy costs. The higher the energy costs, the better the financial argument to make that capital investment.

But phasing out direct combustion is also an opportunity to do things differently, on the household and on the community scale. There's no reason to do a one-to-one replacement of petroleum-derived heating sources with electrical heating sources, since electricity lends itself to a very different set of uses. For example, induction burners are more efficient for cooking than traditional electrical burners in the same way that microwaves are—targeting the energy to a particular material (metal for induction, water molecules for microwaves) means that less energy is wasted in warming up the kitchen. We can replace furnaces with heat pumps or with thermal energy storage, in which hot-water heaters with insulated reservoirs can use inexpensive or excess electricity to heat water and store it for later use, including in radiators or underfloor heating. Or indeed, the opposite: instead of time-shifting electricity use by storing energy in the form of heat, it can be stored in the form of cold. Ice storage air-conditioning is already commonly used in large buildings and college campuses, where ice is produced overnight and water is circulated through it to produce chilled coolant during the day, so that electricity usage for air-conditioning no longer peaks with afternoon temperatures. At the community scale, the hyperlocal nature of sustainable energy can be used to advantage, including using it directly for district heating, skipping the electricity-generation step. In volcanic Reykjavik, homes are geothermally heated by pumping water down to the hotter subsurface regions and then circulating it through a municipal grid. At the other end of the temperature scale, Toronto is building out a deepwater cooling system on the shore of Lake Ontario, pumping cold water up from the depths and circulating it through buildings in the downtown core to keep them comfortable on hot summer days. Moving away from combustion at the point of use opens the door to other ways of meeting basic human needs for food and shelter from the elements.

A Future for Transportation

As an urban resident in the Global North, I get to use an astonishing array of passenger vehicles regularly: my car, trains (subway, commuter, and intercity), buses, and streetcars/trams, plus I take ferries and planes a few times a year. I'm also a cyclist as well as a pedestrian. If our criterion is only that we want our children and grandchildren to have the same sort of access to personal mobility as I do—that all of these specific systems remain functional a century from now—it means that the energy they use needs to be from sustainable sources, ones that don't contribute to greenhouse gas emissions. A simple starting point is to imagine all of the same vehicles, powered by electricity instead of combustion. What might that look like?

Some of them already can or do run on electricity. Vehicles that travel on fixed routes can sidestep the need to carry their own fuel by being connected to the electrical grid instead. It's common for urban public transit, with a third rail that powers subway trains or overhead catenary lines for trams and streetcars. The Amtrak rail line was fully electrified up to Boston in preparation for the launch of the high-speed Acela service that runs through New York to Washington, DC. (Most of Amtrak's passenger service uses dual-power diesel/electric locomotives, a hybrid model that lets them run on non-electrified tracks on more lightly traveled routes and take advantage of electricity when it's available, especially in facilities like New York's Penn Station, where it's preferred because it minimizes pollution and fire risk.)

For vehicles that don't run on fixed routes and therefore do need to carry their own power source, the second approach to electrification is the use of batteries, as in electric cars. While batteries are heavier than gasoline or diesel, there's some serious energy upside. Even the most efficient internal combustion engines in modern cars max out at less than 35 percent useful energy harvested from the heat produced by burning fuel, while converting electricity into motion is more like 75–80 percent efficient. Electric vehicles are so comparatively efficient that no matter where the electricity to charge the batteries comes from, even if it's from coal-fired generating stations, they still use less energy and produce fewer emissions than conventional vehicles. What's more, a conventional vehicle locks in emissions for the entire lifetime of the vehicle, whereas electric vehicles, without any change in how they're used, can actually get cleaner with time as the electricity grid gets a greater and greater proportion of energy from nonpolluting sources. That means there's no reason to wait until the grid is fully decarbonized to transition to electric vehicles.

This is especially true for freight. Heavy trucks are 1 percent of vehicles but account for 25 percent of emissions. Converting trucking fleets to electric vehicles does mean addressing all the same issues as passenger vehicles, including range anxiety and standardizing charging infrastructure. But as these networks get built out, the utility of electric freight vehicles will increase, with significant and continuing cost savings on fuel.

If passenger cars are the most straightforward to decarbonize, flying might be the hardest technical challenge. Powered flight, where every gram matters, is really built around the availability and incredible energy density of petroleum. The high weight-to-energy ratio of even the best commercially available batteries means that electrifying powered flight is still a thorny technical challenge. Some of the most promising technologies for decarbonizing aviation are based around using hydrogen and hydrogen derivatives as fuel instead of batteries, taking advantage of the energy density of chemical bonds without the greenhouse gas emissions of carbon-based fuels.

Transportation infrastructure, particularly for personal mobility, is paradoxical. On the one hand, it can feel like we're making individual decisions about mobility—deciding to have a car, deciding to drive or take the train, deciding to bike or take the bus. On the other hand, perhaps more than any other infrastructural network, the available options are shaped by what's immediately around us: the intersection of the natural environment, the built environment, and all of the human-created elements of transportation systems, from the price of parking to the frequency of the subway to the specific bus route. Once those systems are in place—or removed, as in the case of the streetcar lines that once dominated urban transit in cities like Los Angeles—it's hard to even consider alternatives. Having spent most of my teen years and then twenties as a student in a large city, I didn't learn how to drive until I was nearing the end of my PhD program and wasn't sure where I'd move next. Only after I accepted a faculty position at Olin College did I actually get my driver's license. For my first month as a professor, I took the commuter rail out to the suburbs and then walked forty minutes each way to campus. I used my first paycheck to buy a car, just in time for classes to start. I'd never in my life been alone at the wheel until I picked my new car up at the dealership. My commute was now expensive and infinitely more stressful—a brand-new driver in Boston rush-hour traffic every day!—but it also took less than an hour of my day instead of nearly four. This is what we mean when we talk about much of the U.S. and Canada being car-centric: in many places, the dominant transportation mode is private vehicles because that's what the local environment supports. It's an infrastructure trap. The resources that have been sunk into a specific way of doing something to make it easier means that it's that much harder to do anything else. The alternatives have become challenging or impossible, paradoxically reducing agency and choice rather than increasing it.

Making private vehicles quiet and nonpolluting would be great—don't get me wrong. In many places, especially rural areas, this is likely the path to follow. But once we begin to entertain the possibility of replacing every gas- or diesel-powered vehicle on the road, we also open up new possibilities. Nothing says that we have to build out the same systems and networks we had before—what might we want to change? What new patterns of housing and communities would support different transportation options? If our goal is for our children and grandchildren to have the same access to personal mobility as we do, what might we want to do *differently*?

Replacing every car on the road with an electric equivalent doesn't address all of the *other* issues with private cars, especially in cities, where they dominate neighborhoods by excluding other users of the space, and are a major safety issue. Cars don't scale well. All the negative externalities associated with cars increase as the number or density of cars increases, and there are few, if any, upsides. Cars don't get any cheaper if more people have them, and pollution and congestion only get worse, not better, with more cars.

It's true that I can walk out any time of day or night, get into my car, and drive to nearly any point on the continent, as long as it isn't rush hour and the point in question isn't in downtown Boston. But this car-mediated level of agency is inaccessible to many people: it excludes people who are unable to drive safely because of age (which includes not just older people but children as well) or health, it requires the full and careful attention of a driver, and not only is owning a private vehicle exclusionary simply because it's expensive, but even just firing up the engine means energy costs. While semi- or fully autonomous cars are intended to address some of these issues, they won't address issues of cost or traffic. As someone who drives in congested and idiosyncratic Boston, where we routinely do things like go down the centerline of a two-lane one-way street after a big winter storm because the plowed-aside snow fills the curb half of each lane, I think it'll be a while, if ever, before self-driving cars are tenable in dense, complex, multiuse settings like cities.

Public transit, on the other hand, does scale with density. Like other forms of networked infrastructure, the more people who use it and the denser the nodes, the more effective and useful it is. The point of transportation infrastructure isn't access to any particular kind of vehicle: it's about personal mobility and agency. Two decades after I first started commuting, a new factor has come into play: when I'm at the wheel of my car, I can only drive. I'd happily trade that commute for an hour or so on a train, taking advantage of my laptop, phone, and data networks to get started on my workday or to catch up with my friends.

So much has changed in the last fifty or so years, since the rise of the automobile and the built environment that supports it. A few years ago, I went with Robinson Meyer to visit the Faroe Islands, in the North Atlantic, about equidistant between northern Scotland, Norway, and Iceland. The tiny archipelago is a semiautonomous territory of Denmark, with a total population of about fifty thousand people, about ten thousand of whom live in the largest city, Tórshavn. We flew into the airport, picked up our rental car, and drove via a modern tunnel to the city. Over the past twenty or so years, the transportation and telecommunications infrastructure of the islands has been built out to support a goal of broadening the islands' economic base and making it possible for inhabitants to fully participate in European and global society and culture without needing to relocate to the "big city" of Tórshavn or the European mainland. Rob and I went to the northernmost part of the island and stood on a beach in the long twilight of the summer solstice in the far north, in a village that was home to only a few dozen Faroese souls. Gazing west with nothing but the North Atlantic between me and Labrador, I noticed there were five bars of signal in the corner of my glowing phone. The next day we took a ferry to Sandoy and our host pointed out a new furniture factory, whose location was made possible by those new communications and transport links.

I think about this model all the time: the way that a combination of transportation infrastructure, decentralized energy generation, and modern telecommunications creates possibilities beyond crowded agglomerations surrounded by neglected hinterlands. What if we lived in a multiscalar archipelago of communities within communities, extending through the suburbs and exurbs and into rural areas? It could be made possible by transportation links, supply chains, and high-bandwidth telecommunications networks that allow for distributed work and collaboration. Options for local travel might include e-bikes and microbuses in addition to cars, with mass transit in denser regions, all connected by high-speed transportation links for medium or longer distance travel. Instead of transportation systems that are focused on cities, on moving daily commuters and resources into and out of a central core—the five-day-a-week rush hour commute is already beginning to feel like a holdover from the past—transport networks could facilitate movement from place to place around a more decentralized map, including "chaining" trips. Personal mobility is what we exercise to be with our family and friends, for leisure and pleasure and care, not just for work.

Some Visions for the Next Century

When I read about what major cities were like a century and a half ago, I always find it a bit shocking—they were gray, crowded, loud, dangerous, and polluted. A tiny fraction of the residents lived lives of comfort; poverty, hunger, lack of adequate shelter, and sickness were rampant and accepted. A hundred years from now, when we look back to the way we live today, could we be equally shocked? I imagined playing host to time-traveling neighbor-hood residents from the past. Now I can imagine being the traveler myself, visiting my home in the future.

In this vision, my neighborhood is noticeably greener than it is now, the main street narrow and the pedestrian and cyclist routes wide. The air is clean and quiet, save for the soft hum of electric buses and delivery vehicles. The oldest subway in the U.S. still runs below my feet, and it's bright and modern and frequent, filled with riders, taking them around the city and to the transit hubs of North and South Stations. High-speed intercity rail whisks passengers to cities up and down the East Coast, to the Midwest and into Canada. Or they're heading to the regional airport and on to communi-ties all over the world, the connections with their friends, families, and col-leagues everywhere fostered and sustained by distributed, decentralized telecommunications networks, like the Internet and its successors. As I walk down the street, I recognize a few antique solar panels and wind turbines that are still in service, but most of the structures that harness and store re-newable energy are either unrecognizable to present-day me or, like the cur-rent systems, sit invisibly inside or underneath what I can see. The buildings and vehicles around me track and predict energy generation, storage, and consumption, nodes on a smart grid that share electricity among themselves so it's reliably available where and when it's needed. These energy, mobility, and communications networks underpin daily life and are also the back-bone of resilient, disaster-tolerant infrastructural systems that help manage the human costs of climate instability. Abundant, locally situated energy means that every community is equally well served for their power needs. When heavy weather comes in, whether howling nor'easters in the winter or megahurricanes coming up the coast in the late summer, the local grids stay up and ensure that there is power to ride out the storm. The greenery around me also serves to absorb that heavy rain, a buffer that averts flooding.

If this future seems boring, that's sort of the point! And this is just one possible vision of one neighborhood—it would look different all over the

U.S. and the world. I sketched it out as a future for infrastructural systems not because it's the most likely, or even the most desirable, but because it's an example of the kinds of resilient, sustainable systems that are possible. The future of infrastructural systems everywhere will be locally situated and dependent. The technologies that might serve as the building blocks of any possible future are being invented and refined by the day. I'm not making predictions or providing answers, but instead opening a door to consider different possibilities for a long-term vision of infrastructural systems.

Just about two centuries ago, access to increasing amounts of energy in the form of fossil fuels began to transform our world by powering collective networks, at first slowly and then unstoppably. The value of these systems— the capacity they provide—is so clear that it's almost invisible. As the beneficiaries of these systems, we now have our opportunity in turn. We can recognize what we have in these systems, take everything we learned during the Industrial Revolution, and rebuild them for the coming generations. Transitioning our infrastructural systems to be fully powered by nonpolluting energy sources will make them not only functional and sustainable but intrinsically more resilient and equitable. Rebuilding them gives us the opportunity to reconsider how well they work, and what we might do differently. What does a transportation network look like in a world with the Internet? What does energy use look like in a world with inexpensive renewables?

Resilient and responsive infrastructural systems will be one of our best community tools for responding to climate change and getting through it with a minimum of human suffering. Functional infrastructure is what makes it possible for people to shelter in place, or to evacuate an affected area. Communication networks make it possible to gather and share information and coordinate community responses. Diverse and decentralized energy generation, distribution, and storage mean that communities can be self-sufficient in the face of disruptions, with power for heating, cooling, light, and cooking, and to support neighboring communities.

Rebuilding infrastructural systems to be functional means that they will be locally specific and sustainable, no longer reliant on a steady stream of gas or oil, with a corresponding outflow of money and pollutants. Harnessing renewable energy effectively means it can be abundant, everywhere, providing the opportunity for everyone, no matter where they live, to have access to the energy that serves, as Amartya Sen described wealth, as a general-purpose tool to live the kinds of lives we have reason to value.

There is a possible future for humanity where we have stabilized our climate, where everyone has the energy and resources that they need to survive and thrive, where we get to connect with each other in myriad ways. I get to use "we" in the best possible way, meaning all of humanity. Running the numbers and realizing exactly how abundant renewable energy is, and then realizing how close we are to being able to harness it—it's an absolute game changer. We're accustomed to thinking about making the transition away from fossil fuels to renewable sources as one that we are doing under duress, making a sacrifice to stave off disaster. But that's not what we're doing. What we're doing is *leveling up*. We—you, me, anyone who is alive today—we have the opportunity to not just live through but contribute to a species-wide transition from struggle to security, from scarcity to abundance. We can be the best possible ancestors to future generations, putting them on a permanent, sustainable path of abundance and thriving. And we can do it for all of our descendants—*all* of humanity—not just a narrow line.

But we can only do it together, and we still need to figure out how to get there.

The first big challenge is materials. Rebuilding our infrastructural systems will require an enormous amount of new *stuff*. All those solar panels, wind turbines, and batteries; all those heat pumps, thermal storage systems, new electric stoves and heaters; all those vehicles. The materials cost of a transition like this will be intense because of all the new physical systems that will need to be built. That cost will include everything associated with extracting the necessary raw materials, fabricating the new systems, and transporting them to where they will be used. We've done this once before—we built our modern technological world from scratch, after all—but this transition can't just be treated like an accelerated version of everything that happened over the last century or two, because we know the cost isn't just monetary. It includes all the negative externalities, the environmental and other impacts of both the new systems and the disposal of the now-obsolete technologies. To take this on means fundamentally rethinking how we manage materials throughout their entire life cycle. We live on a planet bathed in sunlight, but it's at the bottom of a gravity well, surrounded by the void of space. For all of human history, we've been living like energy is scarce and matter is infinite, when in fact the opposite is true: we need to learn to live like we have access to unlimited energy, but with the deep understanding that the atoms we have to work with are part of a closed system. Clean, abundant energy will open all kinds of new technological possibilities, but we'll need a deep, society-wide rethink of our relationship to matter, to extraction, consumption, and waste.

The second big challenge is, well, everything else. Knowing that a new set of infrastructural technologies is not just possible but desirable is the necessary first step, but we still have to figure out how to get there. What's changed in the last fifty or so years is that most of the technological challenges have been identified and addressed, and research and development is ongoing, even accelerating. Now most of the remaining questions, and barriers, aren't technological. They are social, political, and economic. Having realized that our infrastructural systems are the main way in which we use and benefit from energy and matter, we understand that we need to transform them. But how on earth do we transform these massive, collective systems? There is no one answer, although the modular, standardized structure of distributed infrastructural systems has the potential to work to our advantage. As with our relationship with materials, we'll need to reframe how we think about these systems.

Ten

RETHINKING THE
ULTRASTRUCTURE

Infrastructural systems make astonishing amounts of energy available in useful form—as light, as heat, as motion, and as generic power to be used as we want, literally empowering us as individuals and as communities, embodied beings interacting with the material world. These systems are the most important practical means we have of collectively implementing any version of "life, liberty, and the pursuit of happiness." That means we can't walk away from these systems, nor can we shut them down and start over.

We also can't continue to do what we've been doing.

Jon Ardern, of the design firm Superflux, wrote that climate change "is not a problem we can 'solve' but rather a predicament we must navigate with responsibility and urgency." Finding a route through the onset of anthropogenic climate change that limits human suffering and ecological damage, means rethinking infrastructural systems. Rather than being reactive, wrong-footed by social and technological failures, we can instead learn how to be responsive to our changing world while building toward something new.

The costs of transforming our technological systems to make them resilient and sustainable are right in front of us, in balance sheets and pro-posed budgets. But the costs and harms of our current infrastructural systems—the ways in which they *don't* work well—are harder for us to see. They shape what we even believe to be possible, and so they make it easy to accept their negative impacts as "just the way things are." Traffic fatalities. Air pollution. A highway that tears apart a neighborhood. Overdrawn groundwater or contaminated water supplies. It's even harder for us to wrap our brains around the benefits of transformed infrastructural systems. It re-quires active imaginative effort to understand what the upsides would be, and the futures they might enable. We have to envision this different world, and choose to work toward it. Every Hurricane Sandy, Maria, or Harvey, ev-ery Camp Fire, every blackout in a summer heatwave or a winter storm, and every multiyear drought is a message from the future where we haven't re-considered our infrastructural systems.

We know that there's such an abundance of energy available on the earth that it's possible to have functional infrastructural systems in fifty years, or a hundred, or more, and we can imagine what that might enable: a resilient, equitable, sustainable future for all of humanity, embedded in a thriving planetary ecosystem. It's a destination to navigate toward. But be-cause these systems generate and sustain our relationships to each other in a particular place, and because they're so contingent, shaped by all the deci-sions that have come before, there really isn't a road map, much less turn-by-turn directions. The geographic, environmental, economic, social, political, and historical contexts of infrastructural systems differ from neighborhood to neighborhood, from continent to continent, and on every geographic scale in between.

For inspiration on ways to think about building new structures for infra-structural provision that better reflect these intertwined goals of sustainabil-ity, resilience, and equitable access, I look to and learn from activists for social change. Prison abolitionist Mariame Kaba, for example, describes the need to act in the short, medium, and long term, and at the individual, community, institutional, and societal levels. But what actions should we take around infrastructure? Building on our understanding of infrastructural systems, where they came from, and how they work, we can pull together a set of heuristics. The six actionable principles for a new ultrastructure that follow can help each of us make decisions about infrastructure in a wide range of different contexts. They are also values to articulate and defend.

1. Plan for Abundant Energy and Finite Materials

We've seen that there is enough energy incident on the planet that it's possible to provide for all of humanity's power needs from renewable sources alone. While there are still challenges, we already have most of the technologies necessary to harness, store, and distribute renewable energy, at all scales and price points, and it's still early in the research, development, and adoption of these systems.

The first heuristic for making decisions about infrastructure is to make this mental and cultural shift and plan for energy abundance. We are not doomed to a dystopian future of failing systems. Responding to climate change is not about making a sacrifice and accepting a more straitened future for ourselves, everyone else on the planet, and generations to come. Responding to climate change is an opportunity: to build out systems that are functional and remain so into the future, and to extend the agency provided by energy and collective infrastructure to communities that, in the past, have had limited access. The proof of concept that humans are capable of building these systems together is all around us. We already live in a world that is defined by the development of technologies and practices that rely on cooperation up to and including the global level—around economics, around standards and systems, around our ecosystems, by the ability to work collectively at all scales.

What does it mean to plan for energy abundance? In the short to medium term, it means spinning up renewable energy systems and transitioning other infrastructural networks like transportation away from fossil fuels, as quickly and completely as possible. By decoupling energy consumption from combustion, greenhouse gas emissions can be reduced without giving up everything that access to energy enables. This doesn't preclude investing in reduced power usage through efficiency, of course. The replacement of incandescent lightbulbs with LEDs over the course of the past decade is an excellent case in point, since LED bulbs use about a tenth of the energy and last about ten times longer. Just as importantly, they produce pleasing light and—especially important in warm climates—much less waste heat. Retrofitting buildings so that they can be heated and cooled more efficiently also makes them more comfortable and secure. Efficiency coupled with abundance isn't about managing scarcity. Instead, it's an investment in thriving and comfort.

This technological transformation would require the wholesale replacement of many of the physical systems in the world, large and small, with entirely different ones. The net result will be a huge amount of embodied energy: all of the energy used to extract the raw materials, process them, manufacture them into what's needed, and deliver them to where they'll be used. Even if it saves years of emissions in the future, the scale of this transformation is all-encompassing, so making all this new equipment with existing combustion-powered systems means digging the climate-change hole we're in even deeper. But the faster that renewable energy sources come online, the faster the emissions cost of making the necessary materials and systems drops. The vicious circle of embodied energy and fossil fuel consumption instead becomes a virtuous circle in which the energy necessary to build out a fully decarbonized infrastructure gets steadily cleaner over time.

Steel and concrete—the basic building blocks of civil engineering—are two of the biggest contributors to greenhouse gas emissions. In part that's simply because they're used in such huge quantities—annual global production is about two billion metric tons of steel and about fifteen billion cubic meters of concrete, or about five hundred pounds of steel and close to three tons concrete per human being on the planet, every year. It's also because they both involve manufacture at high temperatures, in energy-hungry furnaces and kilns that are usually powered by combustion. What's more, the specific chemical reactions that are used to produce both steel and cement (the "active ingredient" in concrete) result in the direct release of carbon dioxide. Emissions-free production will require a transformation in these industrial processes, with all of the research and the investment into new facilities that entails. But for most building projects, the cost of raw materials like steel and concrete is only a small part of the total cost, so getting end users on board early can smooth the transition. The U.S. government, for example, established a "Federal Buy Clean Initiative" in 2022. By guaranteeing a market for carbon-neutral concrete and steel, it incentivizes the switch to zero-emissions production.

Direct emissions and embodied energy are only part of the story. The actual stuff needed to build out these new systems—the copper, aluminum, iron, lithium, cobalt, silicon, polymers, and everything else—has to come from somewhere. There is a significant, and reasonable, fear that the material demands of a transition to renewable energy will lead to yet more extraction, depredation, and conflict, only this time it'll be over lithium and copper instead of oil. Because the earth is a closed system for matter—barring the odd meteorite or spacecraft, nothing comes in or out—every atom also has to end up somewhere. The idea of business as usual—extracting raw materials from the ground, turning them into products, and then dumping the waste as pollution—isn't just politically challenging or morally reprehensible, it's also fundamentally at odds with basic physics. Once we understand that materials are inherently limited, then it follows that we need to figure out how to use and reuse atoms in useful ways, rather than dumping them after they've been used once. Why don't we do this already? Well, the usual reason given is that it's too expensive, that it's not cost-effective to recover and reuse the constituent raw materials once a product has reached the end of its useful lifetime.

From a technological point of view, when we say that it costs too much to do something, it almost always means that it requires too much *energy* to do, and that energy has to be paid for. With cheaper power, even intrinsically energy-intensive processes can be economically viable. Aluminum is a good example. The metal wasn't discovered until 1824, but that wasn't because it's rare—it's the third-most common element in the earth's crust—but because it's such a challenge to purify. It takes an enormous amount of energy to isolate aluminum from its ore and it always will, because it's related to the strength of the chemical bond between aluminum and atoms like oxygen. But the development of a new refining process, coupled with the new availability of abundant and inexpensive energy sources like hydroelectricity, meant that aluminum went from being a precious metal in the nineteenth century to quite literally disposable in the twentieth. This dependence on energy and its cost helps explain why aluminum is one of the few commodities that are currently recycled in quantity, along with steel and glass. All three materials can be melted and remade in processes that need so much less energy than making them in the first place that it offsets the costs associated with collecting and processing—in fact, waste glass (known as "cullet") is essential to making new glass. Plastic, on the other hand, is such a huge and complex set of materials that it's incredibly hard to recycle. To really reuse the atoms in most waste plastic, the polymer needs to be pulled apart into its constituent atoms and then chemically reassembled. Disassembly releases a lot of energy, which is what happens when plastics are incinerated. To reassemble them, all of that energy and more needs to be put back into the material and paid for, which helps explain why plastics are often incinerated or landfilled rather than recycled. Commodity plastics like polyethylene and the plastic used for soda bottles are an important exception to this—they can be mechanically recycled into new products—but the challenge of separating these specific plastics out from the others, particularly in light of the low cost of producing new material, contributes to their dismally low recycling rate.

If fossil fuels are the available source of energy, then the amount needed scales linearly with the amount of energy used, and so the cost of that fuel does too. Energy-intensive processes cost a lot, and the more you make with that process, the more it costs. But with renewable energy sources, the incremental costs of energy can be minimal because there are no longer fuel costs. Electricity produced this way might not quite be too cheap to meter, but it gets close. In a fully decarbonized system where all energy is renewable, this will be true everywhere, all the time. This is a phase change—a fundamental shift in what's possible.

In Ontario, Canada, where I grew up, alcohol sales are highly regulated so most beer in the province is purchased from (where else?) The Beer Store, a retail cooperative for breweries. It mostly comes in standardized brown glass bottles. You take a case of beer home, drink the beer, put the empties back in the box, and return them the next time you go to buy beer. The bottles are cleaned, relabeled, and refilled until they get worn or damaged, at which point they are recycled as cullet and turned into new beer bottles with a recovery rate somewhere upwards of 95 percent. If we create a world where *all* energy is cheap and clean, this kind of circularity can be extended to all of the materials we use, not just the ones where recycling saves energy, and including everything fabricated by putting different materials together. All the plastic we use could be synthesized from waste plastic or from carbon dioxide in the air and then, after its useful lifetime, chemically torn apart and reused to make next week's or next year's plastic products. Instead of mining more and more minerals from the earth, the lithium, cobalt, and other metals in spent batteries would be extracted and upcycled into new ones.

Over the past century we've begun running into the planetary limits of the earth as a closed system. Whether it's the CFCs that damaged the ozone layer or microplastic debris in the deepest oceans, we're learning that we can't just keep discarding waste into the world around us without measurable consequences. Using renewable energy to support reuse and upcycling is an opportunity to move away from "cheaper" extraction, but circular systems of materials usage also minimize the negative impacts of pollution.

This is a cultural change, not just a technological one. Realizing that matter is limited but energy doesn't have to be—both conceptually and in practice—means that the energy cost of recycling and refabrication will no longer be a limiting factor in what's possible. And like beating swords into plowshares, we can imagine transforming all the artifacts of a fossil fuel–powered culture into the ones necessary for a sustainable world. Getting renewable energy fully up and running will change how we deal with physical matter, because access to abundant energy with no fuel costs will make it possible to build out a technological world that's modeled on the natural one, a world where reuse of matter is the default.

In the short term, planning for abundant energy and limited matter might mean conservation and adaptive reuse of existing infrastructure, as well as putting energy and resources into recovering materials or otherwise diverting them from the waste stream. In the medium term, that means continuing research into ways to recover or recycle materials. But in the longer term, we can put the opportunities of renewable energy directly to work on the challenges of limited materials.

Once the cost of energy is no longer a limiting factor, all sorts of otherwise intrinsically energy-intensive processes, like water desalination, become economically and environmentally feasible. I can't imagine everything that readily available renewable energy might make possible, any more than people before the advent of fossil fuels could imagine what my life looks like today, but it would certainly include everyone having access to potable water. In the longest term, we can envision technological systems that are fully aligned with the way that our planet works, with continuous inputs of energy but closed loops for materials. And that system—sustainable, renewable, abundant, where everyone has what they need—is the world that we can work toward. The point of transforming our infrastructural systems isn't just to avoid the worst consequences of anthropogenic climate change—it's to get to this world in which we can use that abundance of energy toward collective thriving, while still respecting the limits and boundaries of our ecosystems and the planet.

2. Design for Resilience

Most of our infrastructural networks are generally designed, operated, and regulated to be robust and reliable, usually through redundancy. But they were built out assuming a stable climate and are implicitly optimized for efficiency—that is, to use the least amount of time, energy, or other resources, which means taking as much slack as possible out of a system. This is especially true for private or investor-owned systems, where optimizing for efficiency is also minimizing costs. We see optimization in the moment-by-moment matching of supply to demand in the energy grid or in the just-in-time nature of global shipping and transportation. In this model, any unused capacity—empty seats on a plane, unused capacity of power plants— is seen as waste. But you can't optimize systems in a context that's changing, especially if it's changing in unpredictable ways. Removing inefficiencies when circumstances are as anticipated means that there isn't much slack in the system to respond when the unanticipated happens. Optimization is intrinsically brittle, because it's about closely matching the output to the conditions, which means it's vulnerable if those conditions change. What we'll need from our infrastructural systems, more and more, is for them to be resilient, able to absorb uncertainty and changing circumstances either without failing or by failing gracefully and reversibly, rather than unexpectedly or catastrophically.

Resilience can take many forms, but one important element of stepping away from just-in-time optimization is building in buffers. A buffer is always having a second box of your children's favorite breakfast cereal in the cupboard for when the first one runs out, and then replacing that back-up box the next time you go grocery shopping. In a world with a changing climate, buffers become a crucial part of the tool kit. An electricity grid that mostly runs on renewable sources will need large-scale energy storage—batteries, pumped-storage hydro, and the new grid-scale storage strategies that are under development—to buffer against the inherently variable nature of renewables. In communities that are susceptible to water scarcity, storage reservoirs act as buffers for seasonal variations and drought. Heavy rainfall runs off impermeable concrete and asphalt, but replacing these materials with soil and vegetation by dechannelizing rivers or creating permeable parking lot surfaces means that water is initially absorbed and then slowly released, or else it sinks down to recharge subsurface aquifers. Either way, the risk of flooding is reduced. Even the San Francisco Bay ferry acts as a buffer, ensuring an alternative to bridges and tunnels in a disaster situation.

A buffer isn't the same thing as a stockpile—it's one extra box of breakfast cereal, not an entire shelf, enough to buy you a few days to get back to the grocery store—but the size of the buffer is related both to how critical a resource it is and to the uncertainty around replenishment. Resilience isn't efficient because it typically requires an upfront or ongoing investment of resources, in the form of time, money, energy, or cupboard space, in order to head off worse outcomes. Humans are flexible and inventive, capable of figuring out ways to respond to changing circumstances. Every time a system fails, whether in a small way or catastrophically, it's humans who have to take up the slack, whether voluntarily or involuntarily. Sometimes the price is low; sometimes it's tragically high. Once you start looking for it, this human-provided slack is everywhere. It might be as minor as the staffer who helps customers at the self-checkout at the grocery store, or leaving home half an hour early to be sure of making an important meeting. It might be giving up a seat on an overbooked flight to take a later one, or it might be as serious as improvising ways to make it through an extended blackout, or the

communities of mutual aid that arise during natural disasters. But being resilient isn't an intrinsic characteristic of humans. It's an active state, and that
means it has a cost in time, agency, labor, or even suffering. Designing infrastructural systems for resilience rather than optimizing them for efficiency
is an epistemological shift as much as anything else. It's recognizing the
uncertainty in the world and preparing to respond to it in a systemic way by
committing the resources to create resilient systems. It means no longer using individual humans, especially the most vulnerable—their time, agency,
safety, and health—as the slack in the system.

3. Build for Flexibility

Many of us are accustomed to the idea of infrastructural facilities being
both big and long-lasting, not just charismatic megastructures like towering
dams and bridges but also busy transportation hubs, huge power plants, and
global telecommunications networks. The size and durability of these systems reflect more than just the scale of provision. Every infrastructural system locks in a way of doing something, and usually that's the point—that
pathway has been made cheaper, easier, more efficient, or more desirable
than the alternatives. It's an approach that makes sense when you have reason to believe that you're building out the right thing in the right way, and
that it'll remain so for a long time. Like optimization, this approach is less
compelling in the face of the uncertainty and change of climate instability,
because making large-scale commitments also boxes out new opportunities
and ways of doing things. We're in the middle of an explosion of new technologies to harness, transform, and distribute energy and materials. Flexible,
exploratory, and contingent models of infrastructural systems are likely to be
a better match for both the uncertainty and the opportunity of the twenty-
first century and beyond.

Maintaining that flexibility also means avoiding making decisions today that lock us into ways of doing into the future. At the individual level,

buying a new car with an internal combustion engine means greenhouse gas emissions for the entire lifespan of the vehicle. But avoiding lock-in—forgoing a further commitment to a now-undesirable system—is genuinely hard, because it's what currently works best. It means pulling yourself out of the sticky entangled spider-web of the existing networks. I have a gas-burning car because, a decade or so ago when it became clear I needed a new one, there wasn't enough charging infrastructure to make relying on an electric vehicle feasible, especially since I'm a "garage orphan," without a dedicated parking space where I could charge it. But the reason I needed a new car *at all* was because of the absence of public transit options.

Every infrastructural system I've described in this book, whether public or private, exists because of specific decisions about the ultrastructure—who pays for what, who owns it, how it's regulated, what negative externalities are permitted. Once it exists, it has the incumbent's advantage. The purpose of every oil and gas distribution system, every pipeline and pumping station, is to make it easier and cheaper to burn fossil fuels, which makes it that much harder to do anything else. It's one of the reasons for strong community opposition to new facilities. If recommitting to existing systems is the path of least resistance—all of the technologies, systems, and standards are already in place to smooth the way—then conversely, transitions to new systems don't just happen. This is where governmental policies at all levels, including regulations and financial subsidies such as grants and tax incentives, have played and continue to play a role. Especially for individuals and households, community education and support can help bridge the gap between the present, where the incumbent options might still have the upper hand in price or convenience, and a future where new options become the new default.

Finally, it doesn't make sense to build out huge new systems intended to last for a long time if the technologies, the systems, and communities they are embedded in are all evolving rapidly. New systems should be smaller scale by default, rather than monolithic, providing opportunities to test-drive systems within particular communities and conditions, and to explore approaches that might be temporary or which serve as a seedbed for more change. In many cities, this included the experimentation with urban space during the first year or so of the global coronavirus pandemic, closing streets to cars and building out additional bike lanes. It might be trial runs of bus routes or on-demand mobility within public transport, like microbuses.

Thinking about infrastructural systems as flexible and exploratory goes against the dominant paradigm of these systems as robust, stable, and big, serving a large area and a commensurately high number of people. But at the same time, the nature of standards and networks means that some important systems are inherently modular, made up of connected parts. The Internet is the canonical example of this, as a network that is decentralized but federated. A "smart grid" for electricity would be a similar model for power: instead of a few large sources (like power plants) and many consumers, it's a model where the network consists of a wide array of electricity generation, consumers, and storage, of varying sizes. The "smart" in "smart grid" is because this approach depends on sensors and computation—a more granular understanding of what the power is doing, and where—not least because of the challenges associated with making decisions within the vastly increased complexity of distributed energy generation and provision.

Infrastructural systems, particularly energy systems, that are designed to make the most of specific local resources to meet local needs can help to sidestep some of the immediate issues around materials extraction and fabrication. A world in which lithium batteries are the only way to store energy looks very different from the one where communities use a wide range of approaches to generate and store power, whether it's gravity batteries, hydroelectric storage, geothermal, or whatever else.

Finally, infrastructural systems that are decentralized but federated embody a model of mutual aid, distinct from top-down provision, and control. The modularization and connectivity of infrastructural systems, together with the inherently distributed and decentralized access to renewable energy, also means that it's possible to begin the transformation of different systems in different ways all over and all at once. This heuristic doesn't only apply to the technological aspects of infrastructural systems. Flexible, distributed infrastructural systems suggest a similar diffusion of political and decision-making power, devolving it to smaller communities. Resisting the consolidation of smaller systems into larger ones pushes back on the concentration of resources, capital, and control. What it means to be small and flexible, or exploratory, or temporary, or decentralized will vary according to the specific situation, but this heuristic is about actively making decisions that favor localization, distribution of resources, and cooperation.

4. Move Toward an Ethics of Care

Nearly the entire history of infrastructural systems is one of using land and resources to create benefits for groups of people, of unprecedented comfort and agency, and simultaneously one of diffusing and displacing harms and negative consequences to other communities and places, of sacrifice zones and organized abandonment. The animating principle behind this pattern has been utilitarianism born of real or perceived scarcity—that meeting the needs and desires of the many outweighs the costs borne by a few—coupled with an accounting of those costs that was colored by racism and othering. In these accountings, Indigenous or African American communities were less valuable than predominantly White communities, the role of colonies was to support infrastructural provision in the metropole, and health risks were more acceptable for lower-income communities than for wealthier ones. When the U.S. Supreme Court ruled that the Tuscarora Nation would lose part of their land to a reservoir, the court meant to ensure that the millions of residents of New York City had a steady supply of hydroelectric power—a utilitarian decision.

Two centuries after the birth of modern infrastructural systems, we recognize that they are a key component of the way that we take care of each other, and we can reject the idea that some lives and ways of life are more worthy than others. Instead of utilitarianism, then, we can draw on a different operative moral framework: the ethics of care. Rather than the abstract idea of maximizing good, caring is a result of compassion and relationships. Ethical care means being attentive and responsive to what's needed, recognizing that those needs vary; it requires taking on responsibility to participate in that care and doing so in a competent way; and it requires communication, trust, and respect. Thinking about infrastructural systems in these terms is fundamentally different from the accounting-sheet approach of utilitarianism. Like infrastructure itself, an ethics of care is collective and relational, which also makes it consistent with infrastructure as a public good, not as private provision or individual responsibility.

Building on an ethics of care rather than utilitarianism as a heuristic for making collective decisions about infrastructure has immediate, practical applications. An ethics of care approach lends itself to reducing inequality because it recognizes and responds to a range of needs, whether it's at the scale of a neighborhood, a country, or the planet. It rightly puts the emphasis on the most vulnerable and marginalized communities that historically have been the most underserved or have borne a disproportionate share of the costs associated with infrastructure. Imagine if we decided that any infrastructural provision that's noisy, noxious, or even just unsightly had to be built in the wealthiest neighborhoods. How would that change what we build? We know the answer to this: the harms would be mitigated.

Thinking about infrastructure through this lens of caregiving also puts operations and maintenance, not innovation, at the center. All of these systems only continue to function reliably because they are themselves cared for, with time, energy, skills, and resources keeping them running. The transformation of infrastructural systems means there'll be no shortage of ribbon-cutting ceremonies for new facilities, but we also need to recognize, plan, and budget for their ongoing care, especially the necessary human labor.

5. Recognize, Prioritize, and Defend Nonmonetary Benefits

It's impossible to put a dollar value on clean water, or electricity, or even telecommunications or transportation. We are each more than our economic capacity and we all have the right to receive the benefits of full societal and civic participation, regardless of whether or not we can produce value as workers. And infrastructural systems, as network monopolies, are not market goods, much less luxuries that we can defer or decide against purchasing. On the provision side, wholesale electricity costs soar during periods of high demand like heatwaves or winter freezes because everyone involved understands that avoiding the true cost of blackouts, in suffering and excess deaths, is worth nearly any price. But without the preparation, skills, and resources to act in the physical world, no amount of money can unstick a frozen gas valve or reboot a transformer damaged by a solar storm. This tells us that it's not enough to think about the costs, benefits, and operations of infrastructural systems only in economic terms. We need to accept this up front and act accordingly, which means being able to recognize, articulate, prioritize, and defend the nonmonetary benefits of these systems.

Many infrastructural systems are public goods in both the economic and everyday sense of the phrase, loaded with positive externalities and nonmonetary value, and we all benefit if these systems are genuinely recognized and treated as what they are. Because all infrastructural systems are embedded in and draw from the world around them, any conception of these systems as public goods needs to go hand in hand with mitigating harms to other public goods and common-pool resources, including environments and ecosystems from the local to the global.

Defending the public good extends into the future. In my required engineering economics course, I was taught how to calculate the time value of money, quantifying how it's worth the most if it's in your bank account today and less and less the further away it is. My colleague Alison told me about learning, as a PhD student in civil engineering, that the Bureau of Reclamation had plans in the 1960s to dam part of the Grand Canyon to make hydroelectricity. The logic was purely financial: the sale of power would cover the cost of construction and then be a source of revenue, and foreseeable profits in the near future would be preferable to the uncertain or limited economic benefits of an empty canyon. With a strictly economic viewpoint, leaving anything on the table today—any resources unextracted—is a lost opportunity. To capitalism, sustainability always looks like underutilization. But the plan to dam the Grand Canyon was on the syllabus so the students could see the limitations of *only* doing the cost-benefit analysis. What actually happened, Alison said, was that "conservationists put up a hell of a fight." Led by the Sierra Club, which ran full-page ads in newspapers including *The New York Times* and *Los Angeles Times* with headlines like IF THEY CAN TURN THE GRAND CANYON INTO A "CASH REGISTER," IS ANY NATIONAL PARK SAFE? YOU KNOW THE ANSWER, the campaign roused huge public support for continuing to protect the region in perpetuity, and the Grand Canyon dam was scrapped.

Nonmonetary costs and benefits can be highly asymmetric. Entropy increases; the universe tends toward disorder. From an energy and materials point of view, it's almost always easier to avoid damaging air, water, and ecosystems than it is to remediate them after the fact. Unpolluting is a lot harder than polluting! Even though natural systems can be surprisingly resilient, recovering rapidly once humans leave them alone, that doesn't mean that they aren't damaged in the meantime. I'm optimistic that humans will get better at restoring ecosystems, including one day even taking on incredibly energy-intensive, long, slow processes like sieving microplastics out of the oceans. But replaced, rebuilt, or remediated is not the same thing as never damaged in the first place, and some damage, like the extinction of species, is likely to be genuinely irrecoverable. Articulating nonmonetary costs is part of how to get to mitigating harms rather than just considering them to be acceptable collateral damage.

A heuristic, then, is refusing to let money set the entire terms for the discussion and decision-making around infrastructural systems, and instead voicing the entire broader context—nonmonetary benefits and harms, positive and negative externalities including environmental impacts, and a consideration of how these extend into the future.

6. Make It Public

Networked infrastructural systems that meet human needs, whether survival or social, are fundamentally incompatible with private investment and profit. As natural monopolies, they're rife with the potential for abuse since their customers are a captive market who always need access to their "product" and who don't have the power to individually negotiate the price, quality, or other characteristics of provision. Since the earliest days of networked infrastructure, investor-held companies have wanted to be in the game because it promised guaranteed revenue at a minimum, and usually also the ability to maximize profit by raising prices. It's why many early private providers were folded into public systems, and one of the main reasons why utilities, whether public or private, are heavily regulated (which means, of course, that the cost of oversight is transferred to the public purse).

Infrastructural systems also rely on common-pool resources. Throughout human history, communities have successfully managed their commons through communication and cooperation, but infrastructure providers have every incentive to exploit these resources for short-term private profit, both through resource extraction and by sloughing the nonmonetary costs and negative externalities of this systems off to the larger population. And they have. We've heard this story over and over again in the history of infrastructure and through to the present day. Most pollution falls into this category: the contamination of public water supplies with chemicals, air pollution around refineries and roads, the ubiquity of plastic waste, or countless other examples, even light pollution in the night sky from fleets of low Earth orbit communication satellites. This misalignment of incentives also means that no amount of regulation and oversight can fully compensate for the continuous pressure, insidious if not obvious, to prioritize activities that will provide profits for investors and to minimize costs that don't have an obvious return, which includes maintenance. The economic impact of externalities and failures falls on users and taxpayers, but because these are infrastructural systems, those risks and harms are more than just financial—they encompass the life, health, and well-being of you, me, everyone we know, and, with pollution and greenhouse gases, all living creatures on our planet.

The issues with infrastructure-for-profit become even more profound if we look toward what it will take to have functioning infrastructural systems for the next century or so. To begin with, no corporation has any interest in the well-being of our descendants twenty or fifty or a hundred years from now, even as potential customers, because the profit motive is fundamentally incompatible with long-term thinking. Investor-owned companies have no interest in creating nonmonetary benefits or positive externalities, other than incidentally, and they are incentivized to ignore negative externalities except when legally required to address them. Profit-seeking is at odds with maintenance and reliability in the present, and it'll be incompatible with resilience and sustainability in the future. To take an obvious example: private power utilities have no intrinsic incentive to sell *less* power, which means that they have little interest in efficiency, much less in distributed renewable energy generation like rooftop solar. The same things that make renewable energy so promising for us as a species—that it's both abundant and intrinsically cheaper than energy from fossil fuels—will make new facilities less and less profitable to build. We're already starting to see this: abundant solar power has begun to result in cheap electricity, which is considered to be a problem for private investment (because that energy can't be sold for a profit) instead of an incredible win for consumers and for society at large.

Well-crafted and -enforced regulations are intended to align corporate behavior with the interests of humans, but the last half century or so of privatized power, water, and telecommunications, in the U.S. and elsewhere, has made clear that they are simply no match for the corrosive incentives of profit. So rather than using regulations as an arm's-length tool of corporate oversight, essential infrastructural systems should be funded and managed as community-owned systems. If they can only be made profitable through the exploitation of consumers, then they should instead be treated as what they are, an investment in the public good. If they're only profitable because the costs of environmental, social, and other harms are externalized rather than considered and addressed, including those resulting from insufficient maintenance, then those "profits" need to go toward mitigating those harms instead of to shareholders. For all the manifold ways that public infrastructure has fallen short in the past two centuries, there is at least a trajectory toward incorporating these broader ideals. Only community-led networks, whether publicly owned or non-profit cooperatives, even have the potential to incorporate broad-based accountability, long-term thinking, and an ethos of meeting needs. And if it's possible to have a well-functioning utility provider that serves all residents equitably, that's embedded in a healthy environment, and still turns a profit? All the more reason for it to be public.

Transforming infrastructural systems to be resilient and sustainable is possible, but it won't be easy. It'll require the build-out of new facilities and systems, and will involve the development of new technologies, all of which will require an investment of money and resources. Technological progress is often thought of as spontaneous and unstoppable, as new technologies emerge in response to unfilled needs or unconsummated desires. But there is almost always one specific criterion that's met, even if it's implicit, and it's this: Someone will make a return on their investment. The new technology will be profitable. Communities including the Amish and Mennonites are often portrayed as anti-technology, but that's not exactly true. What makes them distinct is that they have an intentionally applied set of criteria for what technologies are adopted, based on the impact to the community. Even the Luddites objected to textile machinery not because it was new but because the textile mills meant that control of worker labor, and the profits it generated, was concentrated in the hands of the factory owners.

If the society-wide transformation of infrastructural systems to make them sustainable and resilient is driven by what rewards private investment, it'll be harder to do and we'll miss out on opportunities. If every personal vehicle were to be replaced with an electric equivalent, for example, it would require a huge investment in new cars and would spur the development of related technologies. Now instead, consider building out (or simply growing) public transportation alternatives to driving. They might require a similar total investment and have a similar impact on the economy, but this investment would be collective rather than individual. Public transport has a host of other benefits—safer roads and mobility options for a broader range of people, including the young, the elderly, and others who might not be able to drive, as well as everyone who might struggle to afford to purchase and maintain a car. But it's not either/or, it's both/and, not least because electric vehicles are subsidized by grants, tax credits, and public charging stations, all of which are another way of collectively investing in a transition, but ones that don't open up those new possibilities for collective benefits. To have infrastructural networks that are resilient, equitable, and sustainable, we need to consider a broader set of criteria for the systems and technologies than the possibility of private profit.

The main difference between investor-owned and community-led infrastructure isn't about the investment itself: it's that the first centralizes the financial benefits—the profits—to a small group and the latter leads to more diffuse benefits. The costs of operating and maintaining infrastructure are an expense, or even a loss, from the point of view of a company balance sheet and its investors, but they aren't to the community—those funds are reinvested in the economy through salaries and supply purchases, in addition to all the benefits of having well-maintained infrastructure. And high-quality infrastructure is what enables economic agency and civic participation. The same is true of mitigating pollution or other environmental impacts, with the added benefits of a cleaner environment, and it's especially true of investments made to safeguard these systems into the future. But these decentralized, diffuse benefits are hard to advocate for, for exactly that reason. They don't lend themselves to lobby groups, paid for by the profits from building out new systems. They're what we, collectively and through our representatives, need to argue for, defend, and protect.

Economists don't agree on much, but there is a general consensus that public infrastructure investments can pay for themselves, because any debts that might result will be offset by the increase in economic capacity. This has always been the argument for public investment in infrastructure: what the people of a community or a country can do—and what companies can do and sell—is enabled, shaped, and limited by the nature and quality of the infrastructural services that are available. And that's only the financial value of infrastructural investment. It doesn't capture the enormous nonmonetary benefits of resilient, sustainable, equitable infrastructural provision. A similar calculation holds true for our response to climate change—the estimated cost of building out systems to transition away from fossil fuel consumption and to safeguard our communities will be much less than the cost of not doing so. Once we recognize that these are investments worth making, there are two immediate corollaries. The first is that there are no isolated nodes. Every community is part of a larger social and economic fabric, and so the local tax base and resources can't be the primary driver of whether or not its residents have access to well-functioning infrastructural systems. The second is that, since these infrastructural systems (and other public and common-pool resources) are what make economic activity possible at all, this needs to be reflected in tax structures and policies, particularly corporate taxation. Those funds need to be reinvested back into these common systems.

In practice, what this means is refusing to let public systems like municipal water supplies be put into private hands. It means working to regain public or community ownership of privatized systems, or it means building out public alternatives to supplant them, like community broadband to replace private Internet providers. Funding and tax incentives should prioritize co-ops and other alternatives, instead of serving as a means to funnel public money into existing companies and simply strengthening their network monopolies. It might mean incorporating handover terms, where after a certain amount of time, a private investment (like a toll highway) becomes publicly owned and operated. It also means avoiding large-scale, capital-intensive facilities in favor of small-scale but federated systems, like microgrids. It means actively working against centralization of systems and capital, and it means building and defending community-led, care-first infrastructural systems. There is absolutely no shortage of meaningful work that needs to be done, and there is no shortage of resources with which to do it. The only questions are who will do it, under what terms, and what the resulting world will look like.

Transforming Infrastructure Is Transforming Culture

These six precepts about infrastructural systems are heuristics, not an algorithm or a checklist, because there never is and never will be a one-size-fits-all solution. They're tools for thinking about infrastructure, ideas of what to be working toward and how, as a guide for decision-making over the wide range of situations found in different geographic, political, economic, and physical contexts. These decisions are as much social as they are technical:

what we spend our money on, who benefits, and who decides. There are often going to be conflicts, and balancing the contradictions is going to be a large part of the work. But we also don't have to try to find the perfect answer; we just need to have some drive toward finding better answers, not worse ones. I'm heartened by the large number of committed activists who are consistently applying this pressure and the smart folks in policy who are figuring out ways to support it.

For our infrastructural systems to be truly sustainable, we need more than the technologies and the resources to build out new systems. We need these new ways of thinking about them. When I see a framework like Kaba's, of working toward large-scale transformation by carrying out actions in the short, medium, and long term and at the individual, institutional, and societal level, I immediately envision it as a three-by-three grid and start figuring out what could be placed in each square. So, for example, when I think about electricity provision in the medium term, I think about how solar panels, home batteries, and electric vehicles provide resilience at the individual or household level. At the regional scale, grid-scale batteries can be used to smooth out variations in supply to match demand. And the community-scale version of this? Local centers of renewable power generation and storage that can serve as community hubs in an emergency.

But one key idea here is that all of these things can happen simultaneously, at all these scales, all over. My graduate work in engineering was in materials science. It's the study of how the physical properties of solid matter result from its underlying structures, from the atomic level on up, including the way those structures interact with energy. We're in contact with materials virtually every moment of our lives and so we have a profound empirical understanding of their behavior, but materials science provides a conceptual framework for understanding why they behave the way they do, and reveals the deep connections that underpin their properties. Why are metals shiny, and why do they conduct electricity? Why do some solids melt when they're heated, some soften, and some burn? Why can we immediately distinguish between clear plastic and glass just by touching them? My long engagement with embodiment, materiality, and energy colors how I think about infrastructural systems. We also interact with them continuously, with a deep understanding of how they work based on a lifetime of experience. Materials and infrastructure are both trans-scalar—the way they behave arises from structures across a wide span of sizes, up to the very large for infrastructure and down to the very small for materials. One of the reasons why teaching introductory materials science is so much fun is because it's like giving my students X-ray vision. I can hand them a paper clip to pull open and bend back and forth, and they can feel the wire getting harder and harder to bend before it eventually breaks. Understanding why this is happening is like "seeing" what the metal atoms are doing inside the wire in response to the forces we are applying with our hands. This particular phenomenon is known as "work hardening," and it occurs when metal is bent or shaped without heating it up. Solid metal, like the paper clip, is made up of tiny grains of ordered atoms, in the same way a snowball is made up of individual tiny water crystals packed together. You might have seen this as "spangle" on exposed air ducts in buildings or the crossbars of a chain-link fence, where the zinc coating of galvanized steel forms large grains on the surface, but otherwise the metal crystals are usually only visible under a microscope, and then only on samples that have been cut, polished, and carefully etched to reveal their internal structure. A piece of metal that is cold-worked—whether hammered, or drawn into wire, or otherwise deformed, like bending the paper clip—retains this microscopic structure, but it invisibly accumulates internal stresses between the constituent atoms, which begin to resist moving around. This resistance is what we feel as we push harder with our fingers to get the wire to bend, and it's why it eventually just breaks. But if that heavily stressed piece of metal is put into a furnace and held at an elevated temperature, something remarkable happens. By providing them with *some* energy, in the form of heat, the atoms gain a tiny bit of freedom to move around. The metal remains solid and from the outside it looks the same But inside, the atoms start to shift to relieve the stress that they're under, just slightly. What happens next is astonishing. All through the metal, the atoms spontaneously nucleate *new* crystal structures. First, tiny dots of seed crystals appear everywhere in the metal, all at once, and then they grow and combine and grow as more and more atoms align themselves with the new configurations, and eventually the new crystals completely replace the original, heavily damaged microscopic grains. From the outside, the metal looks the same as it did before. But inside, the interior has recrystallized and rebuilt itself into a completely new structure, without ever breaking the cohesion that holds it together. And this new pattern is no longer stressed and brittle. Instead, it's strong and resilient.

This recrystallization of cold-worked metal atoms is my mental model for how we can transform and recommit to our infrastructural systems. We all live embedded within global networks, but they've been stressed and damaged and are facing more of the same. We want to rebuild these systems, everywhere and all at once, in a way that repairs that damage and relieves those stresses but without passing through atomization, without falling apart. The way this can happen is by local, distributed changes within the larger systems, keeping the connections between them intact as we reshape the systems and replace them with something new. Rather than a top-down replacement, this means giving individual communities the agency and energy—the tools and resources—to reorganize these systems at a local level. This means figuring out how to invest at all levels, making decisions about what length scales to build new systems on, and figuring out which regulations and legislation will need to be unwound and which new policies will need to be created. This work is underway in many places, but there's an enormous amount yet to be done. It's also going to be a new kind of work for many of us.

In order to have functional, resilient, sustainable infrastructure for ourselves and our descendants, we have to practice our relationship to each other at all length scales. At small scales that might be community—the people you are closest to, people who look like you or talk like you or have values similar to yours. But most of these systems, especially in the U.S., have been built out with the explicit intention of having some groups or places benefit while the costs are borne by other people and places. So there's also an element of solidarity—of recognizing that we are all in sustained relationships with all the people who *aren't* like us, because infrastructural systems are trans-scalar hyperobjects, with roots in the past and branches leading on toward our future. It means learning to stand our common ground and to defend the public good, in the face of not only those who would benefit from its erosion but also those who can't understand the value of defending or working toward something that might not involve direct personal benefit.

That doesn't mean that we have to solve huge, overarching issues with huge, overarching solutions. In fact, we need to *resist* these sorts of solutions, because our landscapes and our technologies are in transition, so we absolutely want to be able to explore, experiment with, and even roll back systems as needed, and to do this in different ways in different places. What it does mean is that we need to be able to see and appreciate our infrastructural systems, to imagine what we want them to be, and to nucleate change wherever we are.

Eleven

INFRASTRUCTURAL
CITIZENSHIP

I went to the Hoover Dam almost by accident. I was in Salt Lake City for a conference, and I decided to take a few days afterward to visit Zion and Bryce Canyon National Parks, both a few hours south of the city. The last afternoon of the conference, I picked up a rental car, loaded my bags, and headed south. There was a wildfire just off Interstate 15 and when I stopped for gas, I took a photo of an American flag, illuminated by the dusky orange sun and against a dark background of smoke. I started to pass signs reading LAKE MEAD RECREATION AREA and realized something I hadn't before: Lake Mead is the reservoir that stretches behind the Hoover Dam, and the signage meant that I might be close enough to visit. When I checked into my hotel outside of Zion National Park that evening, I looked at a map and worked out that I could go visit the dam as a day trip, at least in the very western-U.S. sense—at 175 miles away, it'd clock in at about five and a half hours of driving to get there and back. I booked a tour, put together a bag for the road, set my alarm for five o'clock the next morning, and went to bed.

I'd been to the Hoover Dam decades before, during my very first trip to the American southwest, when a friend had convinced me to come with him to Las Vegas for a few days. We stayed in the pyramid-shaped Luxor Hotel, one of the many carefully designed synthetic casino ecosystems lining the Strip, each alive with the trills and beeps of slot machines and flashing lights. He'd rented a convertible for a day and we drove first to the Hoover Dam and then to the Grand Canyon, away from the sensory overload of the city and out into the vastness of the desert. We didn't have time to do much more than gawk at the dam and then to stand and marvel from the nearest point on the canyon rim. By the end of the day, I knew I wanted to return and, while I'd already had my lifetime's fill of the Strip, that meant that some day I'd be back to Las Vegas, as an entry point to get back out to the Grand Canyon and the Hoover Dam. And at seven that morning, there I was. I stopped at a Venezuelan café for breakfast and headed back out of the city.

I had planned my early start in order to arrive at the dam as soon as they opened to visitors. Even at nine thirty in the morning, the July sun was high and bright, and the desert was golden under a brilliant blue sky. As I drove toward the dam, I noticed the line of electricity transmission towers and was momentarily startled when I arrived at the parking area and saw the nearest pylon—the power lines dive down to the turbines at the base of the dam, so it extends out almost horizontally from the edge of the valley wall and looks for all the world like it's caught in the moment of tumbling into the river. The Hoover Dam itself, familiar from so many movies, was a curving wall rising up from the valley floor, with the blue of Lake Mead behind it and huge spillways on either side. If the reservoir is full to capacity, the spillways prevent it from overflowing by diverting water—each can carry the volume of Niagara Falls—around the dam and to the river below. They've only been used twice, once as a test shortly after they were constructed and once in 1983.

Deep in the interior of the dam, the generators in the turbine room were silent that morning, as the demand for electricity was met elsewhere.

While it was only a distant possibility when the facility was built, ongoing droughts in the Southwest could result in the water level dropping below the intake valves for the turbines, making them unusable. Whether or not Lake Mead becomes a "dead pool," the Colorado River will eventually silt up the reservoir, ending the facility's functional lifetime. But the dam itself—the concrete gravity arch structure in the desert—will last effectively forever.

After the tour, I took the elevator from the base of the dam back up to the visitor areas at the top. In the years since I first visited, I've seen many more infrastructural facilities, so I was newly surprised by just how *fancy* everything is. Everywhere you look there is high Art Deco detailing, in brass and stone. There are clocktowers at each end of the dam to show the time in Arizona and in Nevada, since the state and time-zone borders run along the Colorado River (although the hour hands were in the same position that summer day, because Arizona doesn't observe daylight saving time). Even the public restroom at the Hoover Dam felt like it was in a theater or an opera house, a powder room–type space on the lower level with a glorious, hand-laid terrazzo floor and a spiral staircase leading up to the facilities. It's a very similar aesthetic to the Empire State Building, which had been built just a few years earlier than the dam, but that was a commercial building in a thriving metropolis. The Hoover Dam is, well, a *dam*—a utilitarian structure that impounds water and generates electricity, in the middle of the desert, built by the federal government.

Back outside, the red hand of a tower-mounted thermometer was spun most of the way around its dial, reading a scorching 45 degrees Celsius and 112 degrees Fahrenheit under the blazing desert sun. I walked over to Monument Plaza, my attention drawn by the two winged figures in green-patinaed bronze, each thirty feet high, and read the nearby plaque dedicated to the hundred or so men killed during the dam's construction: "They died to make the desert bloom." Under my feet was another clock, for which the mechanism is the gradually changing orientation of Earth in space. Into the terrace around the sculpture, artist Oskar J. W. Hansen embedded a precise schematic of what the stars and planets in the night sky looked like in 1935, when the dam was opened and dedicated. There's a pointer to Polaris, which, as the North Star, sits at the fixed point in the sky that the rest of the stars appear to revolve around. But that point isn't truly fixed. Because Earth precesses on its north-south axis, the North Pole traces out a circle in space and the star that the axis "points to" changes over time. It's Polaris today, but 4,500 years ago, when the ancient Egyptians built the Great Pyramids, the axis pointed to Thuban, an inconspicuous binary star system. Twelve millennia from now, the North Star will be bright Vega. All three are marked on the star map, making it into a clockface that tells the time within this twenty-six-thousand-year cycle of axial precession. It's a message to the far future, to anyone here in this place and observing the remnants of the dam, even twenty millennia from now. Hansen's art makes the Hoover Dam one of the few pieces of public infrastructure I know of that explicitly recognizes an impact on timescales that go far beyond a single human life. It's an infrastructural facility not just as charismatic mega-engineering but as monument.

Part of why I wasn't impressed by my hotel in Las Vegas was because I'd already been to its inspiration and namesake. I have a set of framed photos that I took when I visited Egypt, including one of the Great Sphinx at Giza and another of the Ramesseum, in Luxor, whose toppled granite figures inspired Percy Bysshe Shelley to write "Ozymandias":

> Nothing beside remains. Round the decay
> Of that colossal wreck, boundless and bare
> The lone and level sands stretch far away.

Less than two hundred miles away from the Hoover Dam is a bristlecone pine tree in the Mojave Desert that germinated before the Great Pyramid was built at Giza, almost five thousand years ago. A hundred or so miles north of that tree is King Clone, a ring of self-cloning creosote bushes estimated to be about twelve thousand years old. On my drive south from Salt Lake City, I passed the town of Richfield, home of "Pando," a grove of nearly fifty thousand quaking aspens, in which the trees are genetically identical to each other and share a single root system. While the earliest trees are long gone, the grove is estimated to be thousands of years old, possibly dating as far back as the last ice age. All of these long-lived organisms are embedded in their larger ecosystems, which endure even as they respond to the weather, to the seasons, to gradual shifts in climate and even geology—to everything, in fact, except for the scale and rapidity of human activity.

We expect our infrastructural systems to endure as well, to be there for us day after day, providing the technological bedrock of our lives. But instead of building out these facilities as monuments like the Hoover Dam, we have the opportunity to think of them as ecosystems, like forests. The exact nature of a forest is specific to the local conditions and resources, from the tropical forests of Borneo to the taiga of the subarctic. And while a forest as a whole is coherent in time and space, with continuity and longevity, any individual element that it comprises—every insect, bird, mammal, plant, tree, and fungus—grows, lives, and dies according to its time. A forest is about the *relationship* between the elements, and the systems and benefits that result are a product of the forest in its entirety. A forest is sustainable because it's self-renewing and doesn't produce waste; whatever falls or dies decays and provides nutrients for new growth.

This is our aspirational model for infrastructure. Like forests, our infrastructural systems now have the potential to be modular, networked, decentralized, responsive, and resilient. Instead of a few monumental installations, they could be distributed and locally specific. And above all, they can be sustainable, cradle-to-cradle systems, using energy inputs to create closed loops for materials rather than to dig linear cycles of extraction and waste ever deeper. There's a theory that the reason why we haven't yet observed any signs of extraterrestrial life is because civilizations are faced with an existential choice: either learn to live sustainably, in harmony with their environment, or die. Being "loud"—sending lots of energy or matter wastefully out into space—is a signature of a linear economy. Sustainable societies are quiet.

The Social Grid, and Care at Scale

It's possible to buy some land, build out solar panels, dig a well, install a graywater recycling system, plant crops—a sort of twenty-first century homesteading. My colleague Ben Linder and his partner, Connie, are carrying out this sort of transformation of their home on a pleasant street in suburban Boston, as a way of increasing their resilience and sustainability at the household level. I know folks who live farther afield, off the electrical grid and without immediate neighbors. But their homes aren't bunkers—far from it. The self-sufficiency of a bunker implies (and is motivated by) social as well as technological disconnection. That untethering might be externally imposed, as by a disaster, or it might be voluntary. In contrast, the self-sufficiency of a home like Ben's is deeply relational and embedded in a social grid, just around a different web of technologies. Whether we're on the grid or off it, we're all reliant on those around us. In part this is because virtually everything we use, from modern medical care to copper wiring, was made by or requires the expertise of many, many people, the result in part of the vastly increased technological complexity of the last century. None of us have the wide-ranging expertise to make the artifacts and systems that we rely on to survive. But even without that, a twenty-first century homestead is on a twenty-first century planet, embedded not only in global networks of resources but also in the landscape, with all the changes that it's absorbed and is continuing to absorb—pollution, climate change, food systems. And it's not possible to wind back the clock.

As long as we have bodies, we'll be in a sustained relationship with the people around us. Even if it's on smaller scales of distance and time, shared infrastructure involves being in an ongoing relationship with other people, one that you can't (or can't easily) walk away from. It takes commitment, negotiation, and compromise to navigate the different things that we might want, and the desires and values that they represent, together with those around us. We've *never* lived socially independent lives—if there is a defining feature of our species, it might be that we transmit culture even without direct body-to-body interaction, by drawing it on a cave wall or beaming it out over the airwaves or putting it in writing to be recopied and shared down the decades or centuries. Being alone is like eating only uncooked food— you can do it and survive, but cooked food is the human universal. Not for nothing have we built out communication and transportations systems to live our lives alongside other people.

The converse is also true: we bear some responsibility for the people with whom we are in sustained relationships. As a Canadian by birth and a naturalized American, this is how I think about citizenship. I don't just get a set of rights. It comes with responsibilities to other Canadians and Americans, no matter where we all live and even if we never interact with each other, just by virtue of our being part of the same polity. But since citizenship also carries with it the right to live indefinitely within national borders, that means our bodies are mostly in spaces where they interact with each other—neighborhoods, cities, watersheds, ecosystems, states, the country, the continent. Especially because our infrastructure is such an important way in which we take care of each other, at scale, it's easy to conflate them, but there are two separate relationships here, one political and one spatial. The spatial relationship has nothing to do with being citizens, or taxpayers, or even long-term residents of a particular place. It's just about being present.

What's more, infrastructural systems shape the relationships that we have with people who will be in these places in the future, and the relationships they might have with each other, just like these systems shape our relationships today. Decisions made, facilities designed, and systems built decades ago enable or constrain our daily lives, by making certain possibilities easier or more likely and others harder or impossible. If Boston's Southwest Expressway had been built fifty years ago, instead of being stopped by community opposition, odds are that I would drive on it to work every day. The decisions that we make about infrastructure today—the systems that we decide are "normal"—will in turn shape the lives and choices of people who are not yet born, our collective future. Whether it's a new pipeline or a mandate that new buildings can't have a natural gas connection, shaping the course of the future is literally what decisions about infrastructural systems are intended to do. Infrastructure is how we collectively dream about our future and then bring it into existence.

Recognizing this responsibility to the future, potentially for many generations, means considering resilience and environmental sustainability. In the U.S. and Canada, only a few kinds of institutions genuinely expect and plan to be around in perpetuity. Churches do, and colleges—there's one up the road from me in Cambridge that's three and a half centuries old and counting, and its English antecedents are pushing a millennium. Even the tiny college where I'm a professor, which graduated its first students in the twenty-first century, plans for the long term. So do countries, even if the modern idea of what it means to be a state is only a couple of centuries old.

This idea of being in an ongoing relationship with others simply by virtue of having bodies that exist in the world and which share common needs is what I think of as "infrastructural citizenship." It's a citizenship that encompasses the people who are in a particular place in the world or connected by networks today, as well as those yet to come. It carries with it the responsibility to sustainably steward common-pool resources, including the environment itself, so that future communities can support themselves and each other so they all can thrive. Infrastructural citizenship is not just care at scale, but care in perpetuity.

Passing Through the Veil of Ignorance

The philosopher John Rawls posited a test for any system of societal organization that you might devise, in the form of a thought experiment. I can suggest a way of ordering society: "I shall be queen of all I survey and everyone will do my sensible, fair-minded, and compassionate will." Rawls's test is this: What kind of society would you devise if you would not only have to live there, but you'd also have to pass through a "veil of ignorance"? That is, you wouldn't know ahead of time what role you'd have in that society, or where you'd live, how wealthy you'd be, your ethnicity, gender, social status, or even what your idea of "the good life" might be. Even if I were sure I'd be an effective and benevolent dictatrix, I'd probably rethink my proposal, since there's no guarantee that I'd be the person who gets that role. But we've all, already, passed through the veil of ignorance. None of us chose the circumstances of our birth. Public infrastructural systems, with universal provision, can be seen as a response to this real-life version of Rawls's veil of ignorance by addressing basic needs and being available to everyone. We tend to think that citizenship is about roots, culture, and identity—about our pasts—but infrastructural citizenship is about our present and our future. Rather than being linked only to our ancestors, it's about recognizing and honoring links to our *descendants*, not just our biological heirs but the heirs to place. What matters in infrastructural citizenship is where our bodies are today, the connections we have to those around us, and the impacts that the choices we make will have on the people who live where we are in the future.

The way we live our lives isn't a mechanism that was designed, built, and set into motion, predetermined as by immutable physical laws. Even on this side of the veil of ignorance we can still redesign our societies. As David Graeber wrote, "the ultimate, hidden truth of the world is that it is something that we make, and could just as easily make differently."

The Pragmatism of Universal Abundance

For all that fossil fuels make possible, they also centralize the control of energy (and along with it, wealth), making it inaccessible to most of the people on the planet. My energy-intensive, technologically enabled twenty-first century lifestyle is recognizably utopian to almost all humans that have ever lived. It's a life of peace, comfort, security, mobility, and self-actualization. But I don't want to be as unusually and enormously privileged as I am—I want *everyone* to have my life, not in its exact details, but in what it makes possible. Because infrastructural citizenship is planetary. It might be rooted in home and neighborhood and community, but all human (and nonhuman) life on the planet is now entangled through these systems and, as with national citizenship and restrictive borders, there is no argument that can be made for why only some people deserve to have infrastructural systems that support basic needs.

For wealthy, developed countries that benefited from extractive or settler colonialism, dedicating resources to a global technological transformation might be seen as altruism, although "reparations" would be more accurate. It's also ruthless, selfish pragmatism, because the single best way to safeguard their way of life is to ensure that everyone has access to it. As Elizabeth Tipton Derieux wrote during World War II: "Tomorrow's peace must be more than the absence of armed conflict. It must be just, creative, and cooperative." Especially in a world of climate instability, "just, creative, and cooperative" is also a good description of tomorrow's resilient, sustainable infrastructure.

For the first time in human history, we are within touching distance of having the technology and resources that we need for that collective, energy-mediated agency to be available to everyone. Leaving combustion behind and transitioning to renewable sources means that energy can become a true public good, with no one excluded from using it, and where using it won't take it away from others. It's true that the difficulty in seeing beyond a scarcity mindset has been exacerbated by those with interests in the current energy system, like oil companies. But it's also challenging simply because a world in which energy abundance is possible, with everything that can follow from that, is unprecedented. We did it. Now we have to believe in it.

Imagining Our Futures

In his novels *The Peripheral* and *Agency*, William Gibson provides a conceptually coherent description of a global future that I think about nearly every day. His characters just call it "the jackpot"—a combination of worldwide climate disasters, ecological catastrophes, pandemics, economic crises, and large-scale sociopolitical unrest that together amounted to a distributed, decentralized, slow-motion apocalypse. The scariest part of his portrayal of civilization-wide collapse is that we are recognizably within it, and have been for decades. We know what it looks like, and what it will look like. It'll be business as usual.

One key word in there is "business." If we're not fully attentive to the nonmonetary externalities of infrastructure, the default basis for decisions about what can and does get built will be economic. Especially if those decisions are based on financial return, it builds in a bias toward waiting for things to finally get so bad that remediation—no matter the cost—is cheaper than inaction. Rich communities will invest in renewable energy and climate mitigation strategies. Poorer communities will see more and more of their days go into just trying to survive, on a seasonal and then a monthly and then a weekly basis, and finally they'll just give up on the options. Life expectancies will fall. The U.S. will become a land of climate refugees, of internally displaced people moving from coastal communities that are drowned or southwestern communities that have run out of water to cities that still have infrastructural provision, however minimal. Away from the cities, the country will be a patchwork of armed fiefdoms. More and more lives will be lost on the southern border, California will close its border to Americans from other states as it did briefly during the Great Depression, and the northern border will become militarized for the first time since the War of 1812. Octavia Butler wrote about this future and if there are few positive scenarios in science fiction, there is no shortage of these dystopian backdrops. What's worse, though, is that the U.S. would be a smaller-scale version of the planet as a whole, because if communities that are rich in resources continue to follow this path, they make it that much harder for others to do anything different.

All of these possibilities are manifestations of a bunker mindset, a denial of the web of mutuality that we're all embedded in, at all scales. It's a failure of social imagination, and it scares me. Because however they're realized, whether literally or metaphorically, the size of a room or the size of a country, if atomized bunkers become the norm, it will become much harder to think and act collectively at all. Universal provision via collective systems is how we provide agency and autonomy to everyone, and limiting the freedoms of those around you is unethical and unsustainable, not least because it's almost always enforced through violence.

I don't want to live in that future. You don't want to live there. We don't want this world for anyone. We want to live in that *other* world, the one where we take everything we've learned about energy, networks, and matter, about technology, standards, and collective systems, and we use it all to make a world where everyone's needs are met. Rather than an urban-rural divide, it means building telecommunications and transportation networks that connect everyone. Rather than letting aging infrastructure continue to decline, we can choose to invest in it now and build it to be resilient in the face of climate change. Rather than weaponizing national borders to keep out those who just want the same infrastructurally supported standard of living, we can learn to act collectively, creating these systems everywhere. The future is not being imposed on us. We can decide what future we want to build. None of us want to abandon our neighbors, to bequeath a decaying country to our children, or to see people imprisoned or killed at our borders. Rather than refusing to face the reality of our resource usage and of climate change, or working to mitigate it just for ourselves and our immediate kin or neighbors, we can embrace the economic and social possibility of a just transition to sustainable and equitable systems, a possibility that's based on the recognition that infrastructure is a public and social good, not a money-making opportunity.

We have to imagine, individually and collectively, the kind of world we want to live in. For this, we can learn lessons from people who've fought for change, everywhere, around the world. Infrastructural systems underpin our way of life, and understanding and engaging with them gives us a powerful lever to shift our future. We already know what we need to do to have reliable, resilient, and adaptable systems for energy and resources, and we'll only learn more as we progress. I know that this better world is—that better worlds, plural, are—possible. We need to choose to build them.

Becoming More Human

If my great-great-grandmother were to see me on an evening, puttering around my apartment, making dinner for myself, doing small household chores or just sitting on the couch reading, she could easily conclude that I was an entirely self-sufficient loner because my daily life looks so different from hers. What wouldn't be immediately visible to her is the way that I'm embedded in networks of mutual care. They are social, with my friends and family; professional, with my students and my colleagues; and civic, political, and economic. Each network includes people right next to me and people on the other side of the world. Even though the source of my water or the energy that powers my lights is invisible from inside my home, these infrastructural systems, with the technologies, organization, and enormous amount of human labor needed to keep them functioning, are also networks of mutual care in which I am inextricably bound.

Just like infrastructure, the human work of caring for other humans is also generally taken for granted, underappreciated, and undervalued. It doesn't scale, it doesn't produce a direct return on investment, and it can't be put off or ignored. Nor can it be easily replaced. Futures researcher Jamais Cascio has described a "pink collar future"—while other forms of human work and labor might be replaced by computers or technology, caregiving, which has historically been gendered, won't be. Care is so dependent on context and on embodied, human interaction that it can't easily be automated, made efficient, or optimized. People are intrinsically adaptive, resilient, responsive, locally situated, and appropriate—exactly what we need from our infrastructural systems in an era of climate instability. We need this community-, society-, and civilization-wide commitment to collective systems of care, a pink-collar future for infrastructure. We especially need these commitments as we navigate the transition that is anthropogenic climate change.

Sociality—to care for, learn from, create alongside, and share our days with each other—is just as much a human need as food and water. We consider mobility to be a human right because it's what enables us to be together, and the explosive rise of telecommunications and the Internet over the past decades is in large part because they facilitate social relationships. Not for nothing did humans start creating new ways to interact once our basic needs were reliably met. Culture, learning, shared jokes and shared sorrow, raising our children, caring for our elderly, and together dreaming of and planning for our futures—these activities are the essence of what it means to be human, both individually and collectively. No matter who you are, no matter where you are, no matter *when* you are—a hundred or a thousand years ago or as many years from now—our relationships with each other, and what we create together, won't ever be replaced, mechanized, or automated away. Collective infrastructural systems make it possible to take care of each other's needs at scale, so we can take care of each other (and be taken care of in turn) as individuals. It's what they're for. It's why we need resilient, sustainable, and globally equitable infrastructural systems, ones that reliably meet our basic biological needs, that increase our abilities and agency through access to energy, and that allow us to develop and foster our social relationships with each other through communications and mobility. Our shared infrastructural future is a commitment to our shared humanity.

Acknowledgments

It seems appropriate that, while writing about the global, networked systems that sustain us, I was myself so deeply sustained by my own global network of people.

Thanks to Wayne Chambliss, Charlie Loyd, Robinson Meyer, Matthew Sniatynski, Bradley Garrett, Justin Pickard, Adam Rothstein, Georgina Voss, Sherri Wasserman, Paul Mison, and Paul Graham Raven for being partners in infrastructural explorations, and for all the discussions that took place as we biked on an aqueduct or followed a channelized river or climbed up a mountain or waited in the dark for a container ship to dock.

I'm fortunate to have friends who are writers and so understood, much better than I did, what it would take to turn ideas from a tangled web of synaptic connections in my brain into a continuous line of words on a page. Tim Carmody, Rahawa Haile, Helen Macdonald, Kio Stark, and Clive Thompson all provided practical advice and moral support, for which I'm grateful.

An entire team of "friendly experts" generously shared their time and expertise. A number of them are my colleagues at Olin College of Engineering, including Sara Hendren, Ben Linder, Caitrin Lynch, Rob Martello, Jessica Townsend, Alison Wood, Callan Bignoli and Maggie Anderson. Alison, Charlie, Tim, and Clive, as well as Stephanie Burt, Hillary Predko, and Matthew Weaver, brought their varied perspectives and backgrounds to bear on manuscript drafts. James Abraham, of SolarArise, shared his time and expertise and arranged for me to visit their Telangana facility. All errors and infelicities are, of course, entirely my own.

Courtney Young, my editor, saw the potential in me and these ideas long before I did, and it's been a pleasure working with her, my agent Lydia Wills, and the team at Riverhead Books to bring them out into the world. A grant from the Alfred P. Sloan Foundation supported the writing of this book, and the Boston Athenæum has been the best of all "third places" to work.

My deepest gratitude to my parents, Saroj and Fakir Chachra, for the upbringing that made this book possible, and to my siblings, Kitty, Vinny, and Aneet, for their support. I'm particularly grateful to Aneet and my sister-in-law Ritu, for the many times they opened their home to me to serve as my Los Angeles writing refuge and base camp for West Coast explorations.

In her essay "Holy Water," Joan Didion remarked that not many people she knew carried their end of the conversation when she wanted to talk about California water infrastructure, but my dear friends listened to me talk endlessly about these systems and met it with their own enthusiasm. With love and thanks to Ethan Marcotte and Liz Galle, Sidney Omelon, Scott Purdy, John Pittman, Andrew and Anindita Sempere, John Tinmouth and Tom Chang, and the International Cat Club (including Natalie Kane, Alex Leitch, Lauren O'Hara, and Mols Sauter), and most especially to Rob Ray for all of our adventures, collaborations, and countless conversations.

Finally, my undergraduate engineering students are a source of unfailing enthusiasm and curiosity, and because of them, I never lose sight of the future. To you, and to my niece and nephews Anya, Rayaan, Nikesh, and Aakash, my thanks for the continuing inspiration to think about how we can together build a better world, for everyone.

June 2023

Cambridge, Massachusetts

Further Exploration

Wait, what is that?" Once you begin to see the infrastructural systems in the world around you, there is no going back. Fortunately, there are a number of wonderful visual guides, loaded with photographs, diagrams, and explanations, to help make sense of what's around us. Brian Hayes's *Infrastructure: A Guide to the Industrial Landscape* (W. W. Norton, 2014) is a visual atlas and explainer, with photographs of many of the physical systems in our environment, together with explanatory text. Kate Ascher's books *The Works: Anatomy of a City* (2005), *The Heights: Anatomy of a Skyscraper* (2011), and *The Way to Go: Moving by Sea, Land, and Air* (2014, all from Penguin) have wonderfully informative diagrams. Artist and researcher Ingrid Burrington's *Networks of New York: An Illustrated Field Guide to Urban Internet Infrastructure* (Melville House, 2016) is a hand-illustrated explainer of these newest telecommunications systems.

For more of the stories behind infrastructural systems, I suggest Roman Mars and Kurt Kohlstedt's *The 99% Invisible City: A Field Guide to the Hidden World of Everyday Design* (Dey Street Books, 2020), a companion to their *99% Invisible* podcast. *Places Journal* (placesjournal.org) describes itself as "public scholarship on architecture, landscape, and urbanism," and infrastructure often sits at the intersection of all three. The work there is both accessible and deeply researched, covering the full range of the technology systems in our lives. I'd recommend any of Shannon Mattern's pieces to serve as a starting point. The Center for Land Use Interpretation (clui.org) carries out multidisciplinary research and education projects investigating the human uses of landscape in the U.S., including infrastructural systems, and maintains a deep archive of photographs and other work.

Saul Griffith's *Electrify: An Optimist's Playbook for Our Clean Energy Future* (MIT Press, 2021) is exactly what the title suggests, a detailed guide to the practical aspects of making the transition away from fossil fuels.

While not directly cited in the text, Vaclav Smil's *Energy and Civilization: A History* (MIT Press, 2017) and James C. Scott's *Seeing Like a State: How Certain Schemes to Improve the Human Condition Have Failed* (Yale University Press, 1998) provide many of the conceptual underpinnings for how we think about these systems. This book was largely written before Schafran, Smith and Hall's *The Spatial Contract* (Manchester University Press, 2020) crossed my desk, and it was heartening to see that we highlighted many of the same concerns and arrived at many of the same conclusions despite our different starting points and perspectives. Finally, the opening quotation for this book comes from Ursula Franklin's *The Real World of Technology*, the 1989 CBC Massey Lectures, broadcast across Canada as a six-part radio series and then published in book form (the current edition is available from House of Anansi Press). I came across them as an undergraduate student in the same engineering school where Dr. Franklin was professor emerita. It wasn't until I reread them, decades later, that I realized how profoundly they influenced the ways in which I think about technology, education, and care. Although she passed away in 2016, just as I was beginning to conceive of writing this book, the impact of her ideas and values are reflected in these pages.

Notes

In the course of researching this book, I learned what every historian knows: that every event and system has a long, complex, and contingent story about how it came to pass, and nothing that happens is inevitable. I was necessarily selective about what I included, which also meant simplifying and streamlining. I hope that these notes serve as an invitation to learn more about the research and ideas that I drew on to discuss these complex and fascinating systems.

Throughout the book, I've drawn on a number of sources for data on energy usage and emissions, including:

International Energy Agency, "Data and Statistics," iea.org/data-and -statistics

Hannah Ritchie, Max Roser, and Pablo Rosado, "Energy," Our World in Data, ourworldindata.org/energy

Organisation for Economic Co-operation and Development, "Data: Energy," data.oecd.org/energy.htm

United States Environmental Protection Agency, "Greenhouse Gas Emissions and Removals," epa.gov/ghgemissions

U.S. Energy Information Administration, eia.gov

1. Behind the Lights

1 **One—Behind the Lights:** The academic study of infrastructural systems is a thriving field as part of the multidisciplinary field of science and technology studies, and this paper framed much of my thinking about what infrastructure is. Susan Leigh Star, "The Ethnography of Infrastructure," *American Behavioral Scientist* 43 (1999): 337–91.

2 **Joan Didion describe driving:** Joan Didion, "Bureaucrats," in *The White Album* (New York: Farrar, Straus, and Giroux, 2009), 79–85.

4 **The CANDU design:** Wei Shen and Benjamin Rouben, *Fundamentals of CANDU Reactor Physics* (New York: ASME Press, 2021).

5 **a Randel bolt:** Marguerite Holloway, *The Measure of Manhattan: The Tumultous Career and Surprising Legacy of John Randel Jr., Cartographer, Surveyor, Inventor* (New York: W. W. Norton & Company, 2013)

9 **"you must invent the universe":** Carl Sagan, *Cosmos*, episode 9, "The Lives of Stars," written by Ann Druyan, Carl Sagan, and Steven Soter, directed by Adrian Malone, aired November 23, 1980, on PBS.

15 **coined the word "hyperobject":** Timothy Morton, *Hyperobjects: Philosophy and Ecology after the End of the World* (Minneapolis: University of Minnesota Press, 2013).

15 **Helen introduced me:** Jason Mark, "In Conversation with Helen Macdonald," *Sierra*, June 8, 2016, sierraclub.org/sierra/helen-macdonald-interview-transcript.

15 **Richard Mabey's idea:** Richard Mabey, "Richard Mabey on the Art of Giving Species Their Common Names," *The Guardian*, June 11, 2011, theguardian.com/environment/2011/jun/11/richard-mabey-name-a-species.

16 **highly gendered nature:** See, e.g., "Facts and Figures," UN Women, accessed January 26, 2023, unwomen.org/en/news/in-focus/commission-on-the-status-of-women-2012/facts-and-figures; "Gender Day Gives Rise to the Forgotten Energy Providers in the Climate Conversation," *World Economic Forum*, January 13, 2022, weforum.org/agenda/2022/01/women-biofuel-climate.

18 **warmest years ever recorded:** "2021 One of the Seven Warmest Years on Record, WMO Consolidated Data Shows," *World Meteorological Organization*, January 19, 2022, public.wmo.int/en/media/press-release/2021-one-of-seven-warmest-years-record-wmo-consolidated-data-shows.

18 **major *contributor* to climate change:** "Sources of Greenhouse Gas Emissions," United States Environmental Protection Agency, accessed February 4, 2023, epa.gov/ghgemissions/sources-greenhouse-gas-emissions.

2. Infrastructure as Agency

24 **"cyborg collective":** Paul Graham Raven, "(Re)narrating the Societal Cyborg: A Definition of Infrastructure, and Interrogation of Integration," *People, Place, and Policy* 11, no. 1 (July 28, 2017): 51–64.

25 **the natural ice trade:** Gavin Weightman, *The Frozen Water Trade: How Ice from New England Lakes Kept the World Cool* (London: HarperCollins, 2002)

25 **price per lumen-hour:** Roger Fouquet and Peter J. G. Pearson, "The Long Run Demand for Lighting: Elasticities and Rebound Effects in Different Phases of Economic Development," *Economics of Energy & Environmental Policy* 1, no. 1 (January 2012): 83–100.

26 **proxy for economic growth:** J. Vernon Henderson, Adam Storeygard, and David N. Weil, "Measuring Economic Growth from Outer Space," *American Economic Review* 102, no. 2 (April 2012): 994–1028.

26 **pipelines that were crash-built:** Richard Rhodes, *Energy: A Human History* (New York: Simon & Schuster, 2018), 268.

27 **eating means cooking:** Richard Wrangham, *Catching Fire: How Cooking Made Us Human* (New York: Basic Books, 2009).

29 **the historian Ruth Schwartz Cowan:** Ruth Schwartz Cowan, *More Work for Mother: The Ironies of Household Technology from the Open Hearth to the Microwave* (New York: Basic Books, 1985).

37 **Access to grid electricity:** I only scratch the surface of understanding the electrical grid in this book, but I highly recommend this as a deep dive: Gretchen Bakke, *The Grid: The Fraying Wires Between Americans and Our Energy Future* (New York: Bloomsbury USA, 2016).

39 **per capita power usage:** See note above for sources on energy usage.

3. Living in the Networks

44 **flounder have made a comeback:** Michael Moore, "Tumor-Free Flounder Are Just 1 Dividend from the Cleanup of Boston Harbor," *The Conversation*, January 10, 2019, the conversation.com/tumor-free-flounder-are-just-1-dividend-from-the-cleanup-of-boston -harbor-109217.

48 **the Hohokam people built:** W. Bruce Masse, "The Quest for Subsistence Sufficiency and Civilization in the Sonoran Desert," In: *Chaco & Hohokam: Prehistoric Regional Systems in the American Southwest*, Patricia L. Crown & W. James Judge, eds. (Santa Fe: School of American Research Press, 1991).

49 **the city of Lowell:** I experienced the industrial past of the city as a visitor to the Lowell National Historic Park, created in 1978 as the first site of its kind in the U.S. Cathy Stanton, *The Lowell Experiment: Public History in a Postindustrial City* (Amherst and Boston: University of Massachusetts Press, 2006).

51 **New York's demand for water:** Gerard T. Koeppel, *Water for Gotham: A History* (Princeton University Press: Princeton, 2000).

52 **take the polluted wastewater away:** Sam Bass Warner Jr., *Streetcar Suburbs: The Process of Growth in Boston, 1870–1900*, 2nd ed. (Cambridge: Harvard University Press, 1978), 29–32.

54 **the words of Paulo Freire:** Miles Horton and Paulo Freire, *We Make the Road by Walking: Conversations on Education and Social Change* (Philadelphia: Temple University Press, 1990).

56 **mostly ran on a broad gauge:** George Rogers Taylor and Irene D. Neu, *The American Railroad Network, 1861–1890* (Champaign: University of Illinois Press, 1956).

57 **immediate aftermath of the storm:** Caitlin Dickerson, "Stranded by Maria, Puerto Ricans Get Creative to Survive," *The New York Times*, October 16, 2017, nytimes.com/ 2017/10/16/us/hurricane-maria-puerto-rico-stranded.html.

62 **this web of social structures:** Deb Chachra and Charlie Loyd, "Metafoundry 35: Dilution of Precision," *Metafoundry*, May 11, 2015, tinyletter.com/metafoundry/letters/meta foundry-35-dilution-of-precision.

4. Cooperation on a Global Scale

67 **the intermodal shipping container:** Marc Levinson, *The Box: How the Shipping Container Made the World Smaller and the World Economy Bigger* (Princeton: Princeton University Press, 2006).

68 **happens via standards:** This is a wonderful short primer: Andrew Russell and Lee Vinsel, "The Joy of Standards," *The New York Times*, February 16, 2019, nytimes .com/2019/02/16/opinion/sunday/standardization.html. For more detail: JoAnne Yates and Craig N. Murphy, *Engineering Rules: Global Standard Setting since 1880* (Baltimore: Johns Hopkins University Press, 2009).

69 **universal standard time had gone global:** Vanessa Ogle, *The Global Transformation of Time, 1870–1950* (Cambridge: Harvard University Press, 2015).

69 **the American Revolution ended:** John Keay, *The Great Arc: The Dramatic Tale of How India Was Mapped and Everest Was Named* (New York: HarperCollins, 2000).

70 **Global Positioning System:** Greg Milner, *Pinpoint: How GPS Is Changing Technology, Culture, and Our Minds* (New York: W. W. Norton, 2016).

71 **standards implemented in my laptop:** Brad Biddle, Andrew White, and Sean Woods, "How Many Standards in a Laptop? (And Other Empirical Questions)," preprint, last modified October 22, 2013, papers.ssrn.com/sol3/papers.cfm?abstract_id=1619440.

72 **one for my cup of tea:** *Tea—Preparation of Liquor for Use in Sensory Tests*, ISO 3130:2019 (Geneva: International Organization for Standardization, December 2019), iso.org/standard/73224.html.

74 **all technology has three elements:** W. Brian Arthur, *The Nature of Technology: What It Is and How It Evolves* (New York: Free Press, 2009).

76 **Industrially manufactured knitted clothing:** Virginia Postrel, *The Fabric of Civilization: How Textiles Made the World* (New York: Basic Books, 2020).

77 **Franklin describes such a practice:** Ursula M. Franklin, *The Real World of Technology*, CBC Massey Lectures Series, rev. ed. (Toronto: House of Anansi Press, 1999), 5.

78 **no single human on the planet:** For a deep dive into what it might take, check out Thomas Thwaites, *The Toaster Project: Or a Heroic Attempt to Build a Simple Electric Appliance from Scratch* (New York: Princeton Architectural Press, 2011).

80 **the growth in civil aviation:** Sam Howe Verhovek, *Jet Age: The Comet, the 707, and the Race to Shrink the World* (New York: Penguin Group USA, 2010); "Air Transport, Passengers Carried," World Bank, accessed January 31, 2023, data.worldbank.org/indicator /IS.AIR.PSGR?end=2020&start=1970.

82 **have entirely leapfrogged landlines:** Valentina Rotondi, Ridhi Kashyap, Luca Maria Pesando, and Francesco C. Billari, "Leveraging Mobile Phones to Attain Sustainable Development," *Proceedings of the National Academy of Sciences* 117, no. 24 (June 1, 2020): 13413–13420.

83 **"something small and transitory":** Helen Macdonald, "The Numinous Ordinary," in *Vesper Flights* (New York: Grove Press, 2020), 245–58.

83 **has called this the "systemic sublime":** Tim Carmody, "Every Cup of Coffee Is a Spectacle of Logistics," *Kottke.org* (blog), July 31, 2014, kottke.org/14/07/every-cup-of-coffee -is-a-spectacle-of-logistics.

86 **phenomenon known as induced demand:** Kent Hymel, "If You Build It, They Will Drive: Measuring Induced Demand for Vehicle Travel in Urban Areas," *Transport Policy* 76 (April 2019): 57–66.

86 **Joan Didion described her visit:** Joan Didion, "Holy Water," in *The White Album* (New York: Farrar, Straus, and Giroux, 2009), 59–66.

87 **networked, time-dependent systems:** Donella Meadows, *Thinking in Systems*, ed. Diana Wright (White River Junction, VT: Chelsea Green, 2008).

89 **Within minutes, demand had dropped:** Dan Murtaugh and Brian Eckhouse, "A Text Alert May Have Saved California from Power Blackouts," *Bloomberg*, September 7, 2022, bloomberg.com/news/articles/2022-09-07/a-text-alert-may-have-saved-california-from -power-blackouts.

5. The Social Context of Infrastructure

92 **popularly known as Electric Mountain:** Since my visit in May 2019, the private company that owns Dinorwig Power Station no longer offers tours of the facility to the

general public. The visitor center, which had been closed for refurbishment, is now slated for demolition. "Dinorwig Power Station," First Hydro Company, accessed January 31, 2023, fhc.co.uk/en/power-stations/dinorwig-power-station.

96 **Deirdre McCloskey gave it a name:** Deirdre McCloskey, *Bourgeois Equality: How Ideas, Not Capital, Transformed the World* (Chicago: University of Chicago Press, 2016).

97 **accelerated by European colonialism:** Arguably, all of these factors come together in the earliest corporations that allowed anyone to invest (cooperation), reinvested the profits (exponential growth), and maintained their dominance of geographic regions by force—like the East India Company. Nick Robins, *The Corporation That Changed the World: How the East India Company Shaped the Modern Multinational*, 2nd ed. (London: Pluto Press, 2012).

99 **his book *Development as Freedom*:** Amartya Sen, *Development as Freedom* (New York: Alfred A. Knopf, 1999).

100 **"something other than a belly":** Jean-Paul Sartre, "To Be Hungry Already Means That You Want to Be Free," transl. Adrian van den Hoven, *Sartre Studies International* 7, no. 2 (2001): 8–11.

102 **revealed by excess-death analyses:** Andrew Weber and Mose Buchele, "Texas Has an Official Death Count from the 2021 Blackout. The True Toll May Never Be Known," *Texas Standard*, August 5, 2022, texasstandard.org/stories/texas-freeze-winter-storm-2021 -death-count.

102 **died of carbon monoxide poisoning:** Greg Allen, "A New Hurricane Season Brings a New Threat: Carbon Monoxide Poisoning," National Public Radio, June 1, 2021, npr .org/2021/06/01/1000203891/a-new-hurricane-season-brings-a-new-threat-carbon -monoxide-poisoning.

104 **purchased by the city:** Robert M. Stamp (text), *Bright Lights Big City: The History of Electricity in Toronto.* (Toronto Chapter (TAAG), Ontario Association of Archivists: 1991).

106 **increasingly creepy and intrusive forms:** Ethan Zuckerman, "The Internet's Original Sin," *The Atlantic*, August 14, 2014, theatlantic.com/technology/archive/2014/08/adver tising-is-the-internets-original-sin/376041.

108 **access to a public water supply:** "Public Water Systems," Centers for Disease Control and Prevention, last reviewed March 30, 2021, cdc.gov/healthywater/drinking/public.

109 **not in the economic sense:** Food isn't a public good because *land* isn't a public good, and it needs to be grown somewhere. "Why Isn't Food a Public Good," *Policy Innovations Digital Magazine (2006–2016)*, Carnegie Council for Ethics in International Affairs, September 11, 2014, carnegiecouncil.org/media/series/policy-innovations/policy-innovations -digital-magazine-2006-2016-commentary-why-isnt-food-a-public-good.

111 **The 2021 budget:** National Weather Service, *National Weather Service Enterprise Analysis Report*, June 8, 2017.

112 **"meaningless for anyone but a scholar":** Greg Milner, *Pinpoint: How GPS Is Changing Technology, Culture, and Our Minds* (New York: W. W. Norton, 2016), xvi.

113 **taxpaying residents of 1890s Boston:** Warner, *Streetcar Suburbs: The Process of Growth in Boston, 1870–1900*, 30–31.

116 **"the world in which we live":** Amartya Sen, *Development as Freedom* (New York: Alfred A. Knopf, 1999).

117 **a lampshade that matched your sofa:** "Municipal Electricity in Shoreditch: 'More Light, More Power'" Municipal Dreams, December 3, 2013. municipaldreams .wordpress.com/2013/12/03/shoreditch_electricity.

119 **the subtlest of traps:** Neil Gaiman and Jon J. Muth (illustrator), "Exiles," in *The Sandman: The Wake* (New York: DC Comics, 2019), 141.

119 **the rise of universal literacy:** Max Roser and Esteban Ortiz-Ospina, "Literacy," *Our World in Data*, last revised September 20, 2018, ourworldindata.org/literacy.

119 **Finland was the first:** Bobbie Johnson, "Finland Makes Broadband Access a Legal Right," *The Guardian*, October 14, 2009, theguardian.com/technology/2009/oct/14/fin land-broadband.

6. The Political Context of Infrastructure

125 **In the dissenting opinion:** *Tuscarora Nation of Indians v. Power Authority of New York*, 164 F. Supp 107 (W.D.N.Y. 1958), law.justia.com/cases/federal/district-courts/FSupp/164 /107/1457106.

125 **Bazalgette designed and oversaw:** Stephen Halliday, *An Underground Guide to Sewers: or: Down, Through and Out in Paris, London, New York, &c.* (London: Thames & Hudson, 2019).

128 **With the end of British rule:** Yasmin Khan, *The Great Partition: The Making of India and Pakistan* (New Haven, CT: Yale University Press, 2007).

128 **bringing India fully into the British Empire:** Nick Robins, *The Corporation That Changed the World*, 2nd ed. (London: Pluto Press, 2012); Shashi Tharoor, *Inglorious Empire: What the British Did to India* (Minneapolis, MN: Scribe Publications, 2019).

130 **the country gained its independence:** Angus Maddison, *The World Economy: Historical Statistics*, Development Center Studies (Paris: OECD, 2003), 261.

132 **all artifacts have politics:** Langdon Winner, "Do Artifacts Have Politics?" *Daedalus* 109, no. 1 (1980): 121–36.

133 **planning meetings for the Southwest Corridor:** Karilyn Crockett, *People before Highways: Boston Activists, Urban Planners, and a New Movement for City Making* (Amherst: University of Massachusetts Press, 2018), 181.

133 **thousands of oil derricks:** Mark Olalde and Ryan Menezes, "Deserted Oil Wells Haunt Los Angeles with Toxic Fumes and Enormous Cleanup Costs," *Los Angeles Times*, March 6, 2020, latimes.com/environment/story/2020-03-05/deserted-oil-wells-los-angeles-toxic -fumes-cleanup-costs.

134 **the purpose of a system:** Stafford Beer, "What Is Cybernetics?" *Kybernetes* 31, no. 2 (2002): 209–19.

135 **photo of Berlin at night:** Emma Hartley, "How Astronaut Chris Hadfield Showed Berlin's Ongoing Struggle for Unification," *The Guardian*, April 21, 2013, theguardian.com /world/shortcuts/2013/apr/21/astronaut-chris-hadfield-berlin-divide.

135 **citizenship is an inherited privilege:** Gary Gutting and Joseph Carens, "When Immigrants Lose Their Human Rights," *The New York Times*, November 25, 2014, archive .nytimes.com/opinionator.blogs.nytimes.com/2014/11/25/should-immigrants-lose-their -human-rights.

137 **wealthier nations of the world:** Frank Jacobs, "'The West' Is, in Fact, the World's Biggest Gated Community," *Big Think*, October 12, 2019, bigthink.com/strange-maps/walled -world.

141 **urban highways are a well-known example:** Deborah N. Archer, "'White Men's Roads through Black Men's Homes': Advancing Racial Equity through Highway Reconstruction," *Vanderbilt Law Review* 73, no. 5 (2020): 1259–1330.

142 **chain together segments:** Nancy McGuckin and Elaine Murakami, "Examining Trip-Chaining Behavior: Comparison of Travel by Men and Women," *Transportation Research Record: Journal of the Transportation Research Board* 1693, no. 1 (1999): 79–85.

143 **displacing Native American communities:** Marc Reisner, *Cadillac Desert: The American West and Its Disappearing Water* (New York: Viking Press, 1986).

143 The Southwest Corridor in Boston: Crockett, *People before Highways*.

145 "Somebody Else's Problem field": Douglas Adams, *Life, the Universe and Everything* (New York: Harmony Books, 1982).

145 the term "organized abandonment": Ruth Wilson Gilmore, "Ruth Wilson Gilmore Makes the Case for Abolition," interview by Chenjerai Kumanyika, *The Intercept*, June 10, 2020, theintercept.com/2020/06/10/ruth-wilson-gilmore-makes-the-case-for-abolition.

145 there were differential responses: David Rohde, "The Hideous Inequality Exposed by Hurricane Sandy," *The Atlantic*, October 31, 2012, theatlantic.com/business/archive /2012/10/the-hideous-inequality-exposed-by-hurricane-sandy/264337/; Daniel Aldana Cohen and Max Liboiron, "New York's Two Sandys," *Metropolitics*, October 30, 2014, metropolitics.org/New-York-s-Two-Sandys.html.

146 multiday blackout across Texas: Mose Buchele, *The Disconnect: Power, Politics and the Texas Blackout*, KUT/KUTX podcast, National Public Radio, npr.org/podcasts /1004840920/the-disconnect-power-politics-and-the-texas-blackout.

146 not very evenly distributed: William Gibson, "The Science in Science Fiction," *Talk of the Nation, National Public Radio*, first aired October 22, 2018 [11:22], npr.org/2018/10 /22/1067220/the-science-in-science-fiction.

7. How Infrastructure Fails

147 subplot in which attackers: Louise Penny, *Bury Your Dead: A Chief Inspector Gamache Novel* (New York: Minotaur Books, 2010).

147 specialized air squadron: "The Incredible Story of the Dambusters Raid," Imperial War Museums, accessed February 5, 2023, iwm.org.uk/history/the-incredible-story-of-the -dambusters-raid.

147 destroying hydroelectric power stations: Robert Frank Futrell, *The United States Air Force in Korea, 1950–1953* (New York: Duell, Sloan and Pearce, 1961).

148 Metcalfe Transmission Substation: Richard A. Serrano and Evan Halper, "Sophisticated but Low-Tech Power Grid Attack Baffles Authorities," *Los Angeles Times*, February 11, 2014, latimes.com/nation/la-na-grid-attack-20140211-story.html.

148 highlighted as a potential target: Michael Levenson, "Attacks on Electrical Substations Raise Alarm," *The New York Times*, February 4, 2023, nytimes.com/2023/02/04/us/elec trical-substation-attacks-nc-wa.html.

148 a December 2022 attack: Derrick Bryson Taylor and April Rubin, "What to Know about the North Carolina Power Outages," *The New York Times*, last updated December 8, 2022, nytimes.com/2022/12/05/us/north-carolina-power-outage-moore-county.html.

149 since they're mostly squirrels: Katherine Shaver, "The Bushy-Tailed, Nut-Loving Menace Coming After America's Power Grid," *The Washington Post*, December 25, 2015, washingtonpost.com/local/the-bushy-tailed-nut-loving-menace-coming-after-americas -power-grid/2015/12/25/d4b4c2b6-a8db-11e5-9b92-dea7cd4b1a4d_story.html.

149 the whole U.S. electrical grid: Rebecca Smith, "U.S. Risks National Blackout from Small-Scale Attack," *The Wall Street Journal*, March 12, 2014, wsj.com/articles/SB10001 424052702304020104579433670284061220.

149 targeted by the hacker group: David E. Sanger and Nicole Perlroth, "Pipeline Attack Yields Urgent Lessons about U.S. Cybersecurity," *The New York Times*, last updated June 8, 2021, nytimes.com/2021/05/14/us/politics/pipeline-hack.html.

150 the town of Oldsmar: Andy Greenberg, "A Hacker Tried to Poison a Florida City's Water Supply, Officials Say," *Wired*, February 8, 2021, wired.com/story/oldsmar-florida-water -utility-hack.

152 known as the Carrington Event: Sten F. Odenwald, "Bracing for a Solar Superstorm," *Scientific American* 299, no. 2 (August 2008): 80–87.

152   **that 1989 blackout in Quebec:** "March 13, 1989 Geomagnetic Disturbance," North
American Electric Reliability Corporation, accessed February 5, 2023, nerc.com/pa
/Stand/Geomagnetic%20Disturbance%20Resources%20DL/NERC_1989-Quebec
-Disturbance_Report.pdf.

152   **would be a "gray swan" event:** Nassim Nicholas Taleb, *The Black Swan: The Impact of
the Highly Improbable* (New York: Random House, 2007).

156   **they gave U.S. infrastructure:** American Society of Civil Engineers, *2021 Report Card
for America's Infrastructure*, infrastructurereportcard.org.

157   **A few months after that, the Morandi Bridge collapsed:** James Glanz, Gaia Pianigiani, Jeremy White, and Karthik Patanjali, "Genoa Bridge Collapse: The Road to Trag-
edy," *The New York Times*, September 6, 2018, nytimes.com/interactive/2018/09/06
/world/europe/genoa-italy-bridge.html; Gaia Pianigiani, "Poor Maintenance and Construction Flaws Are Cited in Italy Bridge Collapse," *The New York Times*, December 22,
2020, nytimes.com/2020/12/22/world/europe/genoa-bridge-collapse.html.

160   **lost to leaks:** American Society of Civil Engineers, *2021 Report Card for America's Infrastructure*, 34–35.

160   **revealed by post-failure investigation:** "Highway Accident Report: Collapse of I-35W
Highway Bridge Minneapolis, Minnesota August 1, 2007," National Transportation
Safety Board, November 14, 2008.

161   **"red termite" risk:** "TSG IntelBrief: How a Black Swan Became a Red Termite," The Sou-
fan Center, August 3, 2012, thesoufancenter.org/tsg-intelbrief-how-a-black-swan-became
-a-red-termite.

161   **Edenville Dam in Michigan:** Matthew Cappucci and Andrew Freedman, "Michigan
Dams Fail Near Midland; 'Catastrophic' Flooding Underway," *The Washington Post*,
May 20, 2020, washingtonpost.com/weather/2020/05/20/michigan-dams-fail-midland/;
Jason Hayes, *The 2020 Midland County Dam Failure*, Mackinac Center for Public Pol-
icy, August 11, 2020, mackinac.org/27893.

162   **Assad's military strategy:** Annie Sparrow, "Bashar al-Assad Is Waging Biological War—
by Neglect," *Foreign Policy*, October 24, 2018, foreignpolicy.com/2018/10/24/bashar-al
-assad-is-waging-biological-war-by-neglect.

163   **locating buried infrastructure:** Ingrid Burrington, "How to See Invisible Architecture,"
The Atlantic, August 14, 2015, theatlantic.com/technology/archive/2015/08/how-to-see
-invisible-infrastructure/401204.

164   **Deep in the Andean mountains:** Abby Sewell, "This Suspension Bridge Is Made from
Grass," *National Geographic*, August 31, 2018, nationalgeographic.com/travel/article
/inca-grass-rope-bridge-qeswachaka-unesco.

168   **London's first municipal sewer system:** Stephen Halliday, *An Underground Guide to
Sewers: or: Down, Through and Out in Paris, London, New York, &c.* (London: Thames &
Hudson, 2019).

168   **mapping the locations of homes:** Stephen Johnson, *The Ghost Map: The Story of London's Most Terrifying Epidemic—and How It Changed Science, Cities and the Modern
World* (New York: Riverhead Books, 2006).

168   **worst cholera outbreaks in modern history:** Renaud Piarroux, Robert Barrais, Benoit
Faucher, Rachel Haus, Martine Piarroux, Jean Gaudart, Roc Magloire, and Didier
Raoult, "Understanding the Cholera Epidemic, Haiti," *Emerging Infectious Diseases* 17,
no. 7 (July 2011): 1161–1168.

170   **thousands of American communities:** Emily Holden, Caty Enders, Niko Kommenda,
and Vivian Ho, "More Than 25m Drink from the Worst US Water Systems, with Latinos
Most Exposed," *The Guardian*, February 26, 2021, theguardian.com/us-news/2021/feb
/26/worst-us-water-systems-latinos-most-exposed; J. Tom Mueller and Stephen Gasteyer,

"The Widespread and Unjust Drinking Water and Clean Water Crisis in the United States," *Nature Communications*, June 22, 2021, nature.com/articles/s41467-021 -23898-z.

170 **dozens of First Nations communities:** Leyland Cecco, "Dozens of Canada's First Nations Lack Drinking Water: 'Unacceptable in a Country So Rich,'" *The Guardian*, April 30, 2021, theguardian.com/world/2021/apr/30/canada-first-nations-justin-trudeau-drinking -water.

173 **"the number of things he can let alone":** Henry David Thoreau, "Where I Lived, and What I Lived For," in *Walden* (Boston: Ticknor and Fields, 1854).

173 **his book of the same name:** Michael Lewis, *The Fifth Risk* (New York: W. W. Norton, 2018).

177 **these networks of mutual aid:** In the U.S., there are three groups: the Edison Electric Institute for investor-owned utilities, the American Public Power Association, and electricity cooperatives, which are coordinated by the Federal Emergency Management Agency. Frances Robles, "The Lineman Got $63 an Hour. The Utility Was Billed $319 an Hour," *The New York Times*, November 12, 2017, nytimes.com/2017/11/12/us/whitefish -energy-holdings-prepa-hurricane-recovery-corruption-hurricane-recovery-in-puerto -rico.html.

8. Infrastructure and Climate Instability

180 **Michele Wucker's memorable formulation:** Michele Wucker, *The Gray Rhino: How to Recognize and Act on the Obvious Dangers We Ignore* (New York: Macmillan, 2016).

181 **carbon dioxide is a greenhouse gas:** Joseph D. Ortiz and Roland Jackson, "Understanding Eunice Foote's 1856 Experiments: Heat Absorption by Atmospheric Gases," *Notes and Records* 76, no. 1 (March 20, 2022): 67–84.

181 **A familiar visualization:** The warming stripes visualization was created by University of Reading climate researchers Ed Hawkins and Ellie Highwood, who crocheted stripes into a blanket. Ellie Highwood, "The Art of Turning Climate Change Science to a Crochet Blanket," June 15, 2017, blogs.egu.eu/divisions/as/2017/06/15/the-art-of-turning-climate-change-science-to-a-crochet-blanket/; "Climate Stripes," University of Reading, accessed February 5, 2023, reading.ac.uk/planet/climate-resources/climate-stripes.

182 **the hundredth meridian:** Harvey Leifert, "Dividing Line: The Past, Present, and Future of the 100th Meridian," *Earth Magazine*, January 9, 2018, earthmagazine.org/article/dividing-line-past-present-and-future-100th-meridian.

182 **center-pivot irrigation:** Joe Anderson, "How Center Pivot Irrigation Brought the Dust Bowl Back to Life," *Smithsonian*, September 10, 2018, smithsonianmag.com/innovation /how-center-pivot-irrigation-brought-dust-bowl-back-to-life-180970243.

183 **including the massive Ogallala:** Jane Braxton Little, "Saving the Ogallala Aquifer," *Scientific American* 19, no. 1 (March 2009), 32–39.

185 **blackouts in East London:** Javier Blas, "How London Paid a Record Price to Dodge a Blackout," *Bloomberg*, July 25, 2022, bloomberg.com/opinion/articles/2022-07-25/london -s-record-9-724-54-per-megawatt-hour-to-avoid-a-blackout.

185 **is the Thames Barrier:** "Risk and Uncertainty: Calculating the Thames Barrier's Future," *Carbon Brief*, February 25, 2014, carbonbrief.org/risk-and-uncertainty-calculating-the-thames-barriers-future/; *Thames Estuary TE2100 Plan*, UK Environment Agency, November 30, 2022, gov.uk/government/publications/thames-estuary-2100-te2100/thames-estuary-2100-te2100.

186 **issued a boil-water order:** Carlos Anchondo, "Austin Issues City-Wide Boil Water Notice; Calls for Action 'to Avoid Running Out of Water,'" *The Texas Tribune*, October 22, 2018, texastribune.org/2018/10/22/austin-water-boil-water-notice-after-historic-flooding.

187 **the possibility of tipping points:** David I. Armstrong McKay et al. "Exceeding 1.5°C Global Warming Could Trigger Multiple Climate Tipping Points," *Science* 377, no. 6611 (September 9, 2022), doi.org/10.1126/science.abn7950.

188 **about half of the pollutants:** "Sources of Greenhouse Gas Emissions," United States Environmental Protection Agency, last modified August 5, 2022, epa.gov/ghgemissions/sources-greenhouse-gas-emissions.

193 **total greenhouse gas emissions:** Hannah Ritchie, "Climate Change and Flying: What Share of Global CO2 Emissions Come from Aviation?" *Our World in Data*, October 22, 2020, ourworldindata.org/co2-emissions-from-aviation.

194 **emissions in the U.S. dropped sharply:** Brad Plumer, "U.S. Greenhouse Gas Emissions Bounced Back Sharply in 2021," *The New York Times*, January 10, 2022, nytimes.com/2022/01/10/climate/emissions-pandemic-rebound.html.

194 **their pre-pandemic trajectory:** P. Bhanumati, Mark de Haan, and James William Tebrake, "Greenhouse Emissions Rise to Record, Erasing Drop during Pandemic," *IMF Blog*, June 30, 2022, imf.org/en/Blogs/Articles/2022/06/30/greenhouse-emissions-rise-to-record-erasing-drop-during-pandemic.

194 **advertising campaign for British Petroleum:** Mark Kaufman, "The Carbon Footprint Sham," *Mashable*, accessed February 5, 2023, mashable.com/feature/carbon-footprint-pr-campaign-sham.

197 **an associated increase in emissions:** See note on p. 289 for sources on energy usage.

9. An Emerging Future of Infrastructure

209 **community-level microgrid:** Akhtar Hussain, Van-Hai Bui and Hak-Man Kim, "Microgrids as a Resilience Resource and Strategies Used by Microgrids for Enhancing Resilience," *Applied Energy* 240 (April 15, 2019): 56–72.

209 **cooperatives in rural areas:** Morgen Harvey, "Are Microgrids the Answer to Helping Rural Areas Be More Sustainable?," *Sustainable America*, May 10, 2021, sustainableamerica.org/blog/are-microgrids-the-answer-to-helping-rural-areas-be-more-sustainable.

211 **conditions of ongoing uncertainty:** Iris Tien, "Recommendations for Investing in Infrastructure at the Intersection of Resilience, Sustainability, and Equity," *Journal of Infrastructure Systems* 28, no. 2 (June 2022), doi.org/10.1061/(ASCE)IS.1943-555X.0000684.

219 **energy footprint of Canadians:** "Canada," International Energy Agency, accessed February 5, 2023, iea.org/countries/canada; "Energy Consumption in Canada," World Data, accessed February 5, 2023, worlddata.info/america/canada/energy-consumption.php.

221 **manually adjust every single stovetop burner:** William McKibben, "Apartment," *The New Yorker*, March 17, 1986, 43–91.

224 **many smaller installations:** Varun Sivaram, *Taming the Sun: Innovations to Harness Solar Energy and Power the Planet* (Cambridge, MA: MIT Press, 2018).

231 **the utility of electric freight vehicles:** Amina Hamidi and Enrique Meroño, "How to Decarbonize Heavy-Duty Transport and Make It Affordable," *World Economic Forum*, August 16, 2021, weforum.org/agenda/2021/08/how-to-decarbonize-heavy-duty-transport-affordable.

10. Rethinking the Ultrastructure

241 **"navigate with responsibility and urgency":** Alison Killing, "Superflux Interview," interview by Lemma Shehadi, *ICON Magazine*, Winter 2021, 83.

242 **the need to act:** Mariame Kaba, "Whether Darren Wilson Is Indicted or Not, the Entire System Is Guilty," in *We Do This 'Til We Free Us* (Chicago: Haymarket Books, 2021), 54–57.

244 **basic building blocks of civil engineering:** Rebecca Dell, "Volts Podcast: Rebecca Dell on Decarbonizing Heavy Industry," interview by David Roberts, *Volts*, February 11, 2022, audio and transcript, volts.wtf/p/volts-podcast-rebecca-dell-on-decarbonizing# details.

248 **circular systems of materials usage:** William McDonough and Michael Braungart, *Cradle to Cradle: Remaking the Way We Make Things* (New York: North Point Press, 2002).

257 **the Grand Canyon dam was scrapped:** Marc Reisner, *Cadillac Desert: The American West and Its Disappearing Water* (New York: Viking Press, 1986), 283–290; Sierra Club advertisement, *The New York Times*, June 9, 1966, L35.

260 **abundant solar power:** James Temple, "The Lurking Threat to Solar Power's Growth," *MIT Technology Review*, July 14, 2021, technologyreview.com/2021/07/14/1028461/solar-value-deflation-california-climate-change.

11. Infrastructural Citizenship

270 **The Hoover Dam itself:** U.S. Bureau of Reclamation, "Hoover Dam," accessed February 8, 2023, usbr.gov/lc/hooverdam.

272 **embedded a precise schematic:** Alexander Rose, "The 26,000-Year Astronomical Monument Hidden in Plain Sight," *Medium*, January 29, 2019, medium.com/the-long-now-foundation/the-26-000-year-astronomical-monument-hidden-in-plain-sight-9ec13c9d29b5; Oskar J. W. Hansen and U.S. Bureau of Reclamation, *Nevada Safety Island at Boulder Dam*, 1936, technical drawings, accessed at "Hoover Dam Monument Plaza Safety Island," *Internet Archive*, drawings added March 21, 2013, archive.org/details/Safety Island/mode/1up.

273 **the scale and rapidity of human activity:** Rachel Sussman, *The Oldest Living Things in the World* (Chicago: University of Chicago Press, 2014).

277 **a thought experiment:** John Rawls, *A Theory of Justice* (Cambridge, MA: Harvard University Press, 1971).

277 **"just as easily make differently":** David Graeber, *The Utopia of Rules: On Technology, Stupidity, and the Secret Joys of Bureaucracy* (New York: Melville House, 2015), 89.

278 **As Elizabeth Tipton Derieux wrote:** Elizabeth Tipton Derieux, "The Christian Church in Tomorrow's World," *The News and Observer*, October 1, 1944, 10.

279 **His characters just call it "the jackpot":** William Gibson, *The Peripheral* (New York: G. P. Putman's Sons, 2014); William Gibson, *Agency* (New York: Berkley Books, 2020).

280 **Octavia Butler wrote about this future:** Octavia E. Butler, *Parable of the Sower* (New York: Four Walls Eight Windows, 1993); Octavia E. Butler, *Parable of the Talents* (New York: Seven Stories Press, 1998).

282 **a "pink collar future":** Jamais Cascio, "The Pink Collar Future," *Open the Future*, May 1, 2012, openthefuture.com/2012/05/the_pink_collar_future.html.

Index